THE JOHNS HOPKINS UNIVERSITY PRESS
DIRECTOR'S CIRCLE BOOK FOR 2012

The Johns Hopkins University Press gratefully acknowledges members of the 2012 Director's Circle for supporting the publication of works such as *Johnny Appleseed and the American Orchard*.

Anonymous
Gregory S. Aldrete
Dominic and Helen Averza
Alfred and Muriel Berkeley
John and Bonnie Boland
Darlene Bookoff
Jack Goellner and Barbara Lamb
Charles and Elizabeth Hughes
John T. Irwin
John and Kathleen Keane
Mary L. Kelly
X. J. and Dorothy Kennedy
Anders Richter
Guenter B. Risse
Winston and Marilyn Tabb

JOHNNY APPLESEED AND THE AMERICAN ORCHARD

A Cultural History

WILLIAM KERRIGAN

The Johns Hopkins University Press
Baltimore

© 2012 The Johns Hopkins University Press
All rights reserved. Published 2012
Printed in the United States of America

The Johns Hopkins University Press
2715 North Charles Street
Baltimore, Maryland 21218-4363

ISBN 978-1-4214-0729-6

Book Club Edition

Contents

List of Maps and Figures vii
Preface ix

Introduction 1

1 Seeds 7

2 Becoming Johnny Appleseed 36

3 Suckers 66

4 Walking Barefoot to Jerusalem 101

5 To Serve God or Mammon? 127

6 Yankee Saint and the Red Delicious 162

Notes 195
Essay on Sources 217
Index 225

Maps and Figures

MAPS

Northwestern Pennsylvania frontier, 1795–1804	39
Ohio in 1800	69
Johnny Appleseed country, ca. 1810	84
Northwestern Ohio and Fort Wayne, Indiana, ca. 1830	152

FIGURES

John Chapman preparing a seedling nursery	87
John Chapman's barefoot night run	98
A Swedenborgian tract	111
John Chapman's encounter with a traveling minister	125
Three-dollar bill issued by the unchartered Owl Creek Bank	131
The earliest drawing of John Chapman	157
Shell Chemical using the image of Johnny Appleseed	183

Preface

In the spring of 1970 my first-grade class at Lewis Sands Elementary in Chagrin Falls, Ohio, began preparing to put on a pageant on the American story, to be performed for a cafeteria full of parents eagerly wielding their Kodak Instamatic cameras. Most details of that pageant are lost to me today. But I do have some memory of the first day of preparation for the event, when it was time to assign parts. When the teacher asked the boys in the classroom who would like to play the part of Johnny Appleseed, I and one or two other boys eagerly raised our hands. The part went to a small boy named Hal, the class clown, who had a flair for dramatic gestures, and he proved worthy of the part. When the time came for Hal to stride across the stage, he flung his seeds with the appropriate panache.

There were plenty of other parts in a drama about the nation's founding to delight a classroom full of six-year-old boys. There were the coonskin-cap-wearing frontiersmen, who got to shoulder rifles, and the wild and savage Indians, with feather headbands and rubber tomahawks, and there were many enthusiastic volunteers for these roles. The least desirable role, from my perspective, was that of the pioneer husband, riding on his Conestoga wagon beside a sunbonnet-clad pioneer wife. It was my fate to take the reins of a makeshift wagon, improvised with a couple of cafeteria tables and a great deal of cardboard and construction paper. I envied Hal for his Johnny Appleseed part, and I dreaded the humiliation that accompanied my role—sitting beside a girl and pretending we were married, and dealing with the barbs and teasing of the other boys. It was decades before I understood that the role I and my "wife" played were two of the most important in the whole drama, for the story of America's birth as told to generations of schoolkids is at heart the story of the *domestication* of a continent, and there was no greater symbol of the successful domestication of North America than the pioneer family. Johnny Appleseed was a colporteur of that process.

My interest in the legendary Johnny Appleseed lay dormant for more than a quarter century but was reawakened when a job at a small liberal arts college brought me back to Ohio. I began gathering materials for this book not long after I arrived at Muskingum College (now University) in 1997. I began bringing a bicycle and a kayak along on my research trips so that I could explore more closely the landscapes John Chapman wandered. That these explorations could be considered part of my job reminded me of how privileged I was to hold the position of liberal arts college professor, where I had the freedom to engage in such unconventional but intellectually rewarding research strategies.

Early on in my research I was blessed with additional support through the Muskingum Summer Fellowship Program, which provides opportunities for undergraduate students to work with faculty on real research projects. Erin Stevic, my first Muskingum Summer Fellow, did an extraordinary job organizing the mountains of photocopies and scraps of notes I brought back from these trips into a framework that was still intact when I completed the manuscript more than a decade later, and she also took exploratory trips to archives I had not yet visited, returning with reports about what these places held, which proved invaluable to me when I visited those locations later. As I make the last revisions to this project it seems poetically perfect that I am co-teaching a course with my former student, now a colleague and dear friend.

The financial support of several external organizations were also critical in the completion of this book. Urbana University's Swedenborgian-Scholar-in-Residence program enabled me to spend time in the largest collection of Swedenborgian writings in the United States and to mine the archives of the Johnny Appleseed Society that resides on that campus. A weeklong residency at the Filson Historical Society yielded valuable material on the history of orchard agriculture in the Ohio Valley. The opportunity to participate in the SHEAR-Mellon seminar on undergraduate research with my student Ryan Worbs provided me with the time to explore the William Irvine papers in the Pennsylvania Historical Society archives, a collection that greatly enhanced chapter 2 of this work. A National Endowment for the Humanities Summer Writing Fellowship in 2010 provided the critical push needed to turn a few rough chapters into a completed manuscript.

I am grateful to Muskingum University president Anne Steele and the board of trustees for their continued commitment to supporting faculty research. Early financial support for this project came from the Charlotte Green Foundation and Muskingum University, which enabled me to spend time during spring and summer breaks on driving trips to local historical societies in many of the communities, from Massachusetts to Indiana, that claimed a connection to John Chapman. Two semester-long sabbaticals were also devoted to research and writing this book. In the final years of this project, funds made available from the Arthur and Eloise Cole Distinguished Chair in American History covered the expenses of additional trips to distant archives and for many of the books still stacked around the edges of my computer monitor.

I called upon, and received, generous help from countless professionals and volunteers connected to libraries, archives, and historical societies I visited. All earned my deep appreciation. A few deserve special mention: Joe Besecker, Steven Cooley, and Julie McDaniel at Urbana University; Jan Longone at the Clements Library in Ann Arbor; Linda Showalter at the Marietta College Special Collections and Archives; and, especially, Sheila Ellenberger and the amazing staff of the Muskingum University library, who helped me track down some impossibly obscure old books, tracts, and pamphlets.

My thinking about John Chapman had a long time to mature, as I carried out the regular tasks of teaching, serving on committees, and directing a series of student-centered projects unrelated to my interest in apples and Appleseed. One of the joys of teaching at a small liberal arts college is that it forces scholars outside narrow disciplinary communities and into broader cross-disciplinary conversations. Intellectually, this book is a product of the liberal arts college environment, and I have no doubt that it would look much different were it written in the setting of a large research university. I am deeply indebted to my colleagues in the New Concord Rush the Growlers writing group. Jane Varley, Vivian Wagner, Laura Hilton, Jon Hale, and Sandy Tabachnik approached my draft chapters from a variety of academic disciplines and offered astute criticism and support; conversations with Jim Dooley, professor of biology, helped me grasp the science of apple propagation. Special thanks go to Paul Reichardt, for his mentorship. The interest and encouragement of my departmental colleagues, Amy Bosworth, Karen Dunak, Alistair Hattingh, and Tom McGrath is also deeply appreciated.

Beyond Muskingum, several scholars who were influential in my intellectual development deserve mention: A. J. Carlson and Light Cummins at Austin College, Ken Stevens and Frank Reuter at Texas Christian University, and John Shy and J. Mills Thornton at the University of Michigan all played a significant role in shaping my perspectives and approach to American history. The comments of several anonymous reviewers, who read all or parts of the manuscript, were immensely helpful. Greg Wilson also offered a valuable critique. This book could not have been completed without the critical feedback, deep friendship, and steady encouragement of Marty Hershock and Kevin Thornton. Both read most of this book, including early, very rough, drafts and whole chapters that were ultimately discarded. Their insights made this a more reader-friendly manuscript. More importantly, both kept assuring me that this book needed to be written, at critical points when my time and attention had been diverted to other worthy projects. Finally, I am deeply indebted to my editor, Robert J. Brugger, for his encouragement and invaluable advice.

On a personal note, friends and family members have provided much needed encouragement throughout. My parents, James and Patricia Kerrigan, and my many siblings have provided unending encouragement and support, especially my brother Jim Kerrigan, who shares my love of all things apples and has been an enthusiastic supporter of this project from the beginning. My in-laws, Gary and Margaret Walters, have also provided steady encouragement. But my greatest debts are to my wife, Katie, and son, Liam, who lived with this project for over a decade, and patiently tolerated my distracted mind and my long retreats into the study to complete this manuscript. To them I owe a thousand Saturdays of undivided attention.

Johnny Appleseed and the American Orchard

Introduction

Johnny Appleseed is an American legend. Most people first encounter the myth of the wandering apple tree planter in their childhood. No wonder, for this tale is told in over one hundred children's books, and at least one new or reprinted Johnny Appleseed picture book is published every year. He remains a staple in elementary school curricula across the nation.

While the authors of these books strive to place their own creative imprint on the story, the core elements of the myth rarely change: John Chapman was a pious Yankee committed to a life of exceptional simplicity and benevolence. He determined at an early age to devote his life to one purpose—bringing the blessings of apple trees to the new lands in the developing West. His trees brought sweetness to the hard lives of pioneer families and helped sustain them in their labors. Wandering across the West in bare feet and ragged cast-off clothing, sleeping outdoors, and planting apple seeds wherever he went, Johnny Appleseed took pleasure in denying himself the most basic human comforts in order to carry out his mission. He asked for little in exchange for his trees. He might request a modest payment from those who had cash to spare, but from most he would ask simply for old clothing, a simple meal, or, from the truly destitute, nothing at all. He radiated a spirit of peacefulness that was immediately apparent to all who encountered him. Both Indian and white man trusted him completely and saw him as a man with no malice in his heart. Children especially were drawn to him and greeted him enthusiastically. He loved all of God's creatures and was loath to harm any of them. He was a vegetarian. Books often tell of how he felt deep remorse when in an unusual moment of passion he struck a serpent that had just bitten him. He doused a fire and slept in the cold when he discovered that mosquitoes were flying into the flames to their destruction. His commitments to a life of benevolence and self-denial were driven by his love for God, and he shared that love with anyone who was prepared to listen. As he aged, his eccentricities multiplied, but wherever he went, he was welcomed by every family and embraced as a crazy old uncle. Finally, in myth, his energy for planting trees was superhuman. People across the Midwest, and even beyond, claimed every ancient apple tree in their neighborhood to be one of Johnny Appleseed's seedlings. He was St. Frances of Assisi and Santa Claus wrapped into one bundle.

The myth of Johnny Appleseed is a part of our national origin story, in which the United States expands into the trans-Appalachian West in the years after the American Revolution. Johnny Appleseed isn't the only hero in this drama, but he is quite different from its other heroes. Daniel Boone, Davy Crockett, and Mike Fink present a jarring contrast to the gentle tree planter. Violence—directed at Native Americans and nature—lay at the heart of those other myths, while Appleseed is remembered for his extraordinary meekness and generosity, and for sowing, not destroying. The short explanation for this difference is that the folk tales developed at distinct historic moments and thus served different cultural and social needs. The myths of Boone, Crockett, and Fink first flourished in the age of Jackson and reflect that era's obsessions with masculine aggression. The myth of Johnny Appleseed, in contrast, was a product of the Victorian era, when sentimental feeling and feminine traits were more commonly celebrated.

Appleseed, along with Boone, Crockett, and Fink, received updates during the Cold War as each was deployed to serve new concerns. Among the most powerful disseminators of these legends was the Walt Disney Company, which seized on the westward expansion story to target a new audience of baby boomer children.[1] Disney sanitized the most gruesome aspects of the Crockett and Fink traditions, yet even after this cleansing, the contrast with Johnny Appleseed remained startling. Mike Fink, Davy Crockett, and Daniel Boone were archetypes of American manhood, and even in the Disney versions, violence was nearly always central to their stories. Disney also added the thoroughly mythical Paul Bunyan to this cast and celebrated him for his prowess felling whole forests of trees. Johnny Appleseed, in sharp contrast, devoted his life to planting them.

Most American children of the Cold War era understood Johnny Appleseed to be a member of the same team of frontier superheroes despite his difference from the others. Boone, Crockett, Fink, Bunyan, and Chapman were all actors in a drama about transforming a continent. Crockett and Boone cleared the land of menacing Indians and wildlife; Fink helped make the interior rivers safe for commerce; Bunyan cleared the forest; and Appleseed planted fruit trees to prepare the land for white American farm families. In Cold War versions of these stories, Boone and Crockett reluctantly used violence as a last resort. These heroes protected American families from a red menace on television shows like *Walt Disney Presents* and *Daniel Boone* at a time when American soldiers were doing the same in other parts of the world. In that context, Johnny Appleseed symbolized the other American response to the threat, winning hearts and minds with charity and benevolence. If Crockett's war against the Red Stick Creeks explained American military involvement in Korea, Appleseed's unbounded benevolence was a metaphor for another approach to the same danger, manifested in American aid programs and organizations like the Peace Corps.

The myth of Johnny Appleseed is based on a real person. John Chapman's life paralleled the years of the nation's creation. Born in Leominster, Massachusetts, on the eve of the American Revolution, John Chapman had roots that reached back to the early years of English colonization in the region. He died in Fort Wayne, Indiana, less than a year after Samuel Morse transmitted the biblical query "What hath God wrought?" along a telegraph line, signaling the birth of modern communications. Chapman grew up amid significant changes in American life. The Revolution brought with it political, cultural, economic, and even spiritual turmoil for Chapman and his generation, and he matured in a world where ideas like democracy, equality, and pluralism, once deemed undesirable, were ascendant. The Revolution also opened up new land in the West, and this offered Chapman, like thousands of others streaming into Ohio, Indiana, Illinois, and beyond, hope that he could fulfill the dream that had eluded his father. In the trans-Appalachian West, Chapman lived between European and Native American worlds. The Second Great Awakening and the radical perfectionist enthusiasms of the era brought about a spiritual transformation in John Chapman. He aged in a world where direct access to national markets presented the once isolated farmers of Ohio and Indiana with new opportunities and new risks. And while the market advanced their material comfort, like all changes it also fostered different reactions to the new rules of commerce.

While this work attempts to recover the story of John Chapman's life, it is not a conventional biography. It does not assess Chapman's impact on his times, calculate the number of apple trees he planted during his lifetime, or determine the veracity of each specific local claim about a John Chapman nursery. Instead, it seeks to tell the story of John Chapman as a way of illuminating the cultural landscapes he inhabited. In this sense it might fall into that ill-defined genre sometimes called microhistory.[2] John Chapman was not a representative everyman for his time. If he were, he would almost certainly be forgotten today. It was neither his wealth nor his achievements that made him a worthy subject of tall tales. Nor was it only his eccentricity that led to his enduring popularity. There is no escaping the fact that John Chapman was peculiar, with his stubborn devotion to the seedling apple tree, his radical self-imposed poverty, and his unorthodox religious ideas. However, for his neighbors, retelling stories of John's odd behaviors—whether told in a spirit of derision or affection—was also a way of talking about their own concerns.

John Chapman's life and the cultural landscapes he inhabited cannot be understood without telling the story of the apple in America. In modern America, the apple has emerged as the most ordinary and uncontroversial of all fruit. If we associate John Chapman with the modern Red Delicious, that bland staple of the elementary school lunch box, then much of the meaning of his story is lost. That such a fruit could tempt Eve to original sin or be "the apple of discord" that sparked the Trojan War is hard to imagine. The apple's story is part of the nation's story, and that history is more interesting than that glossy, red lunch-box apple might lead one to believe. It was a fruit suffused with cultural meaning, and at times it served as a symbol in contests between Native Americans and European invaders, between poor whites and richer ones, between the tippler and the teetotaler, and between those embracing the modern age and those nostalgic for the past. A few words about the history and science of apple trees are necessary before we proceed further.

The origins of *Malus pumila* or *Malus domestica*—the common apple tree—are in central Asia, possibly in the mountains of Kazakhstan, where locals still harvest apples of great diversity in size, shape, and color from whole forests of wild apple trees that cover the mountainsides.[3] The modern grocery store apple is the result of centuries of cultivation, selection, and hybridization. Still, the seeds of every modern grocery-store apple contain a radical genetic diversity connecting them back to the wild apples of Kazakhstan. Gather one hundred seeds from a favorite apple variety and plant them, and the trees that grow will produce fruit with dramatically varied characteristics, most quite different from the parent apple tree and most unpalatable for fresh eating. Planting apple trees from seed is entering a genetic lottery, but every so often this lottery produces a winner, a fruit of exceptional qualities worth propagating. Many popular apple varieties today were the happy accidents of seedling trees. The simplest way to propagate a specific apple variety is through grafting—attaching a twig or branch of the favored variety to the rootstock or a branch of another apple tree. The fruit that grows beyond that graft is essentially a clone of the fruit on the tree from which it came.

The grafted apple tree is a marvelous example of what human-plant partnerships can achieve. If you are lucky enough to live near a public or pick-your-own orchard today, you can see this by walking among the trees. On older grafted trees, the spot where the rootstock and graft were united can be seen as a small scarred bulge in the trunk just a few inches above the soil. Many characteristics of the tree are determined by its rootstock, but the characteristics of its fruit are largely determined by the graft. Some rootstocks thrive in the harsh winters of the far north; others endure damp low-lying soils with stoic determination. The apple's incredible adaptability and ambitious range, which allow it to thrive from the Carolina piedmont to the Maine backcountry, from New England's rocky soils to the dry valleys of central Washington, is made possible by the diversity of its rootstock.

Perpetuating apple varieties through grafting was a skill that had been practiced for centuries in the Old World before Europeans arrived in America. The result was a great variety of apples of exceptional size and sweetness. In contrast, the native peoples of North America, despite tending to highly cultivated annual crops like maize, generally gathered wild fruit rather than cultivating it. The North American crab apple was part of the diet of some peoples, including the Iroquois, who tempered its bitterness by roasting it in the fire or marinating it in maple syrup. When Englishmen arrived in North America, some continued to keep orchards of grafted trees, but others abandoned grafting and raised orchards of seed-grown trees for reasons that will be explained later.

The distinction between a seedling and grafted apple tree is lost on most Americans today, but it was common knowledge for Americans in the first half of the nineteenth century. In his essay "Wild Apples," Henry David Thoreau celebrated the virtues of the seedling apple tree, which by his time was facing extinction in his neighborhood as Concord, Massachusetts, raced into the modern age. For Thoreau, apples served as a suitable metaphor for the story of the nation. The indigenous crab apple, *Malus coronia*, he compared to the Indian. Both appeared to be equally rare in Massachusetts by his time, as he confessed that he never encountered a native crab apple tree until May 1861. Thoreau had little enthusiasm for the grafted varieties of *Malus pumila*, which he associated with bland uniformity of modern times; instead, he celebrated those grown from seed, trees he called "wild apples," comparing them to the hardy pioneer. Thoreau was a great contrarian, and few of his neighbors shared his declared preference for the gnarled, bitter, wild apple. But most would have appreciated his analogy. For at the heart of the European conquest of the Americas was an agro-ecological revolution, one fueled by the spread of cultivated Old World crops. The Old World apple, Thoreau contended, "emulates man's independence and enterprise. It is not simply carried but, like him . . . has migrated to this New World and is . . . making its way among the aboriginal trees."[4] One need not embrace an ecological determinist narrative to acknowledge the role that the Old World's biological imports—pathogens, most significantly—played in the conquest of North America. But on a more intentional level, Old World agriculture, which transformed landscapes through a mixed husbandry regime that combined the cultivation of both annual and perennial crops with livestock raising, was at the center of the story of that conquest. The Old World apple tree did not have the destructive impact of colonists' hogs unleashed in the woods, but it sometimes played a more subtle role, planting European ideas of property on the landscape.

If the distinctions between *Malus coronia* and *Malus pumila* served in some small way as a symbol of the cultural divide between Indians and Europeans in America, the difference between the seedling apple tree and the grafted one served as a symbol of cultural divides within European America. Thoreau's celebration of the bitter "wild apple" was a part of his greater contrarian campaign against the transformations expanding capitalist markets were making in his time. While Thoreau praised the seedling apple's virtues, agricultural reformers of his age were busily condemning backward pioneer farmers for their persistent attachment to seedling trees. The campaign against the seedling apple tree is a story that has been largely forgotten, but it can help us understand John Chapman and his life. By growing apple trees only from seed and rejecting grafting, Chapman had taken a side in a culture war of his time. Understanding Chapman's close association with the seedling apple tree—Thoreau's "wild apples"—is essential to understanding his life and how he was perceived in his own time.

It was John Chapman's eccentricities that first preserved him in local memory, but later developments transformed those local traditions into a national myth. Along the path from local memory to icon in the national origin story, the meaning of John Chapman's real life, and that of the seedling trees he planted, was obscured. This work seeks to recover those stories from obscurity.

CHAPTER ONE

Seeds

In 1638 a miller from the north of England named Edward Chapman, great-great-grandfather of John "Appleseed" Chapman, sold off his possessions and sailed for New England. Edward was part of a wave of religiously motivated English Puritans who crossed the Atlantic in a "Great Migration" between 1630 and 1642. The costs of such a move were substantial, so the majority of those who participated in this migration were Englishman of middling circumstances, or the servants of such people. Many made the voyage to America as whole family units, but Edward came as a single man. Arriving in Boston on October 2, 1638, it took Edward a few years to get established, but by 1642 he was married, and two years later had attained the status of freeholder in the new town of Ipswich, thirty miles northeast of Boston.[1] Despite his background as a tradesman—something he shared with many other men who joined the great migration—Edward remade himself as a farmer in this new world. Edward may have embraced the opportunity to possess and till a piece of his own land, but it was also a change made by necessity. New England in its early years was abundant in land but scarce in capital and labor. Many skilled tradesmen as a result made the transition to farmer.[2]

AGAWAM

John Winthrop, Jr., the son of the colony's first governor, carved the boundaries of Ipswich from marshy coastal lands, trisected by the Ipswich and Essex Rivers, a region called Agawam ("fish-curing place") by the Sagamore Indians who had come there annually for generations to harvest the abundance of migratory fish. For southern New England's indigenous peoples, Agawam was an important part of their mixed subsistence lifestyle, which employed a strategy of seasonal mobility to glean sustenance from agriculture, hunting, fishing, and gathering.

Cultivated plants, in the form of the three sisters—corn, beans, and squash—were the staples of the Sagamore diet, but they also feasted on game from the winter woods, fish pulled from the rivers in spring, and a variety of nature's wild bounty, including blueberries, strawberries, crab apples, and chestnuts.

This seasonal mobility—going to the food in its due season and trusting that the land would provide for them—was the Sagamore contract with the natural world. Contrary to popular myth, the Sagamore and other tribes of the region did indeed have a concept of property, but it was a property ownership that was collective, not individual, and sometimes not exclusive. Ownership was defined by specific uses, as revealed in the name Agawam—"this is the place where we cure fish." In addition to their annual migration cycles, the Sagamore and other Algonquin peoples of the region followed a practice of long-cycle rotation in agriculture, cultivating plots for eight to ten years, then shifting to new lands, allowing old fields to go fallow and even return to forest for thirty or more years. Southern New England's indigenous peoples appear to have sustained a population density of about five people per square mile using this system, before European diseases began to decimate their numbers.[3]

John Winthrop, Jr., and the small band of English Puritans who accompanied him to Agawam in 1639 had different ways of making a living and different understandings of ownership. The Agawam they entered had no plowed fields, no fences, no apple orchards, no permanent shelters, and for much of the year, it had no residents. The Sagamore might claim it as theirs, but from the young Winthrop's perspective, they were poor stewards of the land. Gazing across the broad flat fields of Agawam, Winthrop imagined a town and named it Ipswich, after the English commercial center from which many of the men in his group had first departed for the New World. As Winthrop's men marked off the boundaries of this future community, they were surely struck by the geographical similarities—Ipswich, England, was also situated on a navigable tidal river—and imagined one day that this land would become a thriving commercial center surrounded by tidy, prosperous English farms. But the name Ipswich also had religious meaning for them. Ipswich had been an early center of reformation Protestantism in England, and between 1538 and 1558, at least six men and three women were burned alive, tied to a stake, for refusing to recant their Protestant convictions. Known collectively as the Ipswich martyrs, their story was told again and again by these later Puritans who had left England to escape religious persecution and establish a holy commonwealth of believers.[4] The founders of the Massachusetts Bay Colony—and the new town of Ipswich—intended to make a *new* England, one that retained many aspects of the world they left behind but that was also more just and righteous; one in which order and deference prevailed but poverty and extreme inequality were banished; and one in which all, or nearly all, shared a common faith.⁵

Edward Chapman was among the earliest settlers of this new Ipswich and played his part in remaking Agawam to fit the Puritan designs. Edward and his neighbors replicated English agriculture to the extent that the new climate and geography allowed them and imposed English concepts of property ownership on old Agawam. The mixed husbandry Edward Chapman and the other new farmers of Ipswich employed departed from Sagamore agriculture in important ways. It employed some different crops; it incorporated domesticated livestock as both food and a source of fertilizer; and it added the cultivation of a perennial food crop, orchard fruit. These new agricultural strategies provided both opportunity and challenges, but they could not coexist with Sagamore ways. Once the English mixed-husbandry system was established, Edward and his neighbors got nearly all of their food from their crops and livestock; hunting and gathering provided only incidental sources for the typical Puritan farmer, and eventually the forests and open lands where these activities occurred were replaced with barns, fields, and orchards. Moreover, this mix of livestock raising and annual and perennial crop cultivation could sustain greater population densities and ensured a lifestyle with a much greater degree of fixity; the dung of their cattle, sheep, and hogs enabled Edward and his neighbors to return nutrients to the soil, prolonging its fertility, increasing their yields, and permitting the land's use indefinitely. Standard apple trees might not begin producing peak crops for ten years, but they could provide an abundance of fruit for decades, helping to foster an investment in place that further encouraged them to stay put. Intensive agricultural practices meant that Edward and his Ipswich neighbors saw no need for periodic relocation, and apple trees, barns, fences, and houses provided plenty of incentives for staying in place.

The first Old World apple trees in Massachusetts Bay Colony actually preceded the great Puritan migration. Clergyman William Blaxton (aka Blackstone), who arrived at the site of Boston as early as 1625, planted an orchard on the slopes of Beacon Hill before the city was founded. Blaxton became legendary for his eccentricities and allegedly rode his trained bull through the streets of Boston, doling out apples and flowers to his new neighbors. By 1635, he had relocated to the more tolerant colony of Rhode Island, where he again was remembered as the bringer of the first apple trees to that state. In early histories of Rhode Island, Blaxton is credited with developing what might have been the first named apple in North America, the Yellow Sweeting, which later gained widespread popularity as the Rhode Island Greening.⁶ Apples and eccentricity appear to be fellow travelers, and one wonders if the young John Chapman heard the stories of Blaxton during his school days and found some inspiration in them.

The English mixed-husbandry regime with its grain crops, livestock, and orchards was not transplanted to Ipswich perfectly. Promoters of the Puritan colony in New England portrayed its climate as similar to that of old England, reinforcing the expectation that the migrants could expect to replicate the best English ways there. In contrast, promoters of the Virginia colony emphasized the climactic *difference* and suggested colonists might earn great profits growing oranges, figs, olives, and other valuable crops that could not be cultivated in England. Both depictions proved to be cases of wishful thinking. The Chesapeake Bay region was still vulnerable to killing frosts, and early experiments with semitropical fruit-growing quickly failed; New England's continental climate was more extreme than England's—winters were longer and more severe, and farmers faced a shorter growing season between last and first frost. In addition, the rocky soils were generally not as productive as the ones they had left behind, and the native grasses their livestock grazed on did not pack the nutritional punch of English grasses. Among the first struggles English farmers faced in their new land was that English barley and wheat were very labor intensive and hard to cultivate in stony fields. While New Englanders never entirely abandoned these favored crops, they did turn to alternatives that were easier to grow in these new conditions, Indian corn and apple trees. While fermented apple cider was a common drink in certain parts of England, most seventeenth-century Englishmen celebrated and sang about "John Barleycorn," the folk god of their preferred beverage, beer. But with barley a scarce commodity, residents of many New England towns switched to drinking apple cider instead.⁷

The increased importance of hard cider was only one way in which New England's mixed husbandry diverged from that of the mother country. The art of grafting to produce trees with fruit of consistent qualities was an ancient one and had been practiced in England for centuries, but most New England farmers opted to raise seed-grown trees instead. This was not for lack of knowledge of the science of grafting but for practical reasons. Edward Chapman and his neighbors lived in a land-abundant, but capital- and labor-scarce world. Shipping grafted fruit stock of favorite English varieties took up valuable cargo space, and keeping that stock alive on a long sea journey was difficult. As a result, any grafted stock arriving from England would fetch a high price. Also, Edward and his neighbors depended on family labor in building a farm and could scarcely afford the time required to graft and tend to expensive apple trees. Furthermore, many of the English apple varieties carried across the ocean did not fare well in the harsher climate of New England. Planting from seed not only preserved scarce capital and labor, but it also enabled farmers to "laboratory test" their trees. Those seedlings that proved their climate hardiness in the farmer's nursery they transplanted to the orchard, planted in a dooryard, or scattered about meadows, while those that withered they discarded. Edward Chapman and his neighbors preferred climate-hardiness over fruit quality or yield, something traditional farmers in premodern societies where food security was paramount have always favored.[8]

Livestock, fences, fields, and orchards defined the Puritan world, shaping Puritan ideas about property and forming the blueprint of their physical vision for the societies they sought to create. This was the framework under which Ipswich's founders operated as they sought to establish order and remake the landscape in their new town. The Massachusetts Bay Colony's General Court granted a sizeable tract of land to Ipswich's founders. The initial grant for Ipswich was bounded to the north by the mouth of the Rowley River, to the east by Ipswich and Squam Bays and Gloucester, and to the south by lands controlled by Salem. This grant, which ran roughly six miles inland, includes today all of the lands not only in the present town of Ipswich but also Topsfield, Hamilton, and Essex. The founders of Ipswich followed the custom of most other new towns and began with a division of only a fraction of the lands granted by the General Court. This "first division," as it was called, determined a settler's economic and social standing, so one of the keys to achieving standing in these new Puritan towns was to arrive early enough to be part of the first land division, and Edward Chapman arrived right on time to benefit.

While the Puritan founders hoped to create a world without the great disparities of wealth and poverty that existed in England, they were by no means egalitarian. Differences in economic and social standing, John Winthrop and other leaders believed, served important purposes. They staved off anarchy and encouraged Christian behavior. Those slightly better off were expected to treat those below them with Christian charity; those with less were expected to show deference to town leaders. John Winthrop, Jr., and the other founders of Ipswich used the first division to establish a social ranking, as did the founders of other towns. Some settlers received larger plots of land; after the first decade of settlement, latecomers were generally excluded from receiving any town lands at all; instead, they were encouraged to go west and find a place in a newer town. Subsequent divisions of land generally reinforced the established hierarchy, with lands being distributed in the same proportions to those present for the first division.[9] The concepts of private property and rank meant that Edward Chapman's Ipswich shared more in common with the England that he had left behind than with the Sagamores' Agawam he helped to erase.

Soon after Edward Chapman arrived, he met and married Mary Symonds, the daughter of another Ipswich landowner. Land ownership gave Edward a competency, the means to economic independence. Back in England, a competency might also be commonly achieved by acquiring the skills and tools of a tradesman, but in the early years of New England, few could earn a living solely by a trade. Land ownership was in the earliest years the path to a competency for the overwhelming majority of New England's men. The equation of land ownership and independence was reinforced by law: property was one of the two requirements a young man like Edward needed to fulfill to become a freeman, a status that gave him a voice in his community, the other requirement being full acceptance into the community's church as a "saint," a person who by his testimony and behavior had convinced the fellow saints that he was predestined for heaven. To become a freeman and saint was the goal of nearly every male who voluntarily joined the great migration, and Edward was able to achieve this status by 1644. At some point in that year, Edward took the following oath at Ipswich's meeting house, in front of all of the other freeman and most of the town's residents:

> I, [Edward Chapman], being by God's providence an inhabitant and freeman within the jurisdiction of this Commonwealth, do freely acknowledge myself to be subject to the government thereof, and therefore do here swear by the great and dreadful name of the everlasting God, that I will be true and faithful to the same, and will accordingly yield assistance and support thereunto, with my person and estate, as in equity I am bound; and I will also truly endeavour to maintain and preserve all the liberties and privileges thereof, submitting myself to the wholesome laws and orders, made and established by the same. And further, that I will not plot nor practise any evil against it, nor consent to any, that shall so do, but will truly discover and reveal the same to lawful authority now here established, for the speedy preventing thereof. Moreover, I do solemnly bind myself in the sight of God, that when I shall be called to give my voice, touching any such matter of this state, wherein freemen are to deal, I will give my vote and suffrage, as I shall judge

in mine own conscience may best conduce and tend to the public weal of the body, without respect of persons or favor of any man; so help me God in the Lord Jesus Christ.

Those adults in Ipswich who had not achieved the status of freeman, by either some economic or spiritual failing, were also required to demonstrate their obedience and submission by taking a "resident's oath." They were required to swear to "accordingly submit my person, family, and estate to be protected, ordered, and governed by the laws and constitutions" of Ipswich and the colony and to promise to alert authorities about any seditious behavior they witnessed, but they were not "called to give [their] voice" on matters of governance as freemen were. For Edward's generation, the presence of non-freemen residents was tolerated but not welcomed, a reality grudgingly accepted because even in this land of abundance and grace, some people would fail to meet the economic and spiritual expectations required to become a full member of the community.[10]

Over the next fifteen years, Edward and Mary Symonds Chapman added five children to their family. After the birth of their fifth, Mary died, perhaps as a result of complications in her last delivery. Edward soon remarried a local widow, Dorothy Swain, but the new union produced no additional children. Getting a living from the stony soils of New England was hard work. As his children aged, Edward grew increasingly dependent on their labor—to attend to farm chores, to improve and fence more land, to raise more livestock, and to put more soil under the plough. Edward Chapman was motivated to do so not for the pursuit of wealth for its own sake but in order to pass along a competency to his sons, so that they might one day achieve standing as freeholders in Ipswich or a nearby town. His ability to be a patriarch and keep his large family in close orbit required him to amass land, a need that resided in tension with John Winthrop's strictures that saints practice Christian charity. How could believers like Edward Chapman reconcile the kind of land-grabbing required to provide for the futures of their large families with their promise to love their neighbors and show charity toward all? Concerns for the family's welfare would win out in this contest; Ipswich's freeholders quickly stopped dispensing additional lands to newcomers, despite the generous boundaries of Ipswich's original land grant. Latecomers had to go elsewhere; second and third divisions of land were mostly reserved for the original families and largely dispensed within the lifetime of this first generation.[11]

The Puritan plantation in New England was designed to be a profitable enterprise, but the founders of the colony also had ambitions to build in this new world a reformed version of the one they had left. At times their aspirations bordered on utopian, but their Calvinism, which constantly reminded them of man's innate depravity, reined in this impulse to some extent. While the communities Edward Chapman and his fellow saints created in this new world resembled those they left behind in many aspects, the differences were nothing short of revolutionary. Among the radical innovations was the abandonment of the practice of primogeniture, the law in England since 1066.

Born of England's demographic reality—too many people sharing a limited amount of land—primogeniture granted the bulk of family property to the eldest male heir in order to ensure that property did not become carved into plots too small to sustain a family. But New England offered a radical new promise— a seemingly unlimited amount of land for farming—allowing the Puritans to abandon primogeniture in favor of the principle of partible inheritance, which divided land among all sons and also usually provided something for daughters.[12] This new practice held out the possibility that a successful patriarch might provide a competency in the form of land to all his sons. But the gap between what was legally possible and what was practically achievable was not easily bridged, and over time, providing for all one's children became more difficult to achieve. Edward Chapman would have to amass enough land in Ipswich to provide for four sons, while his fellow Ipswich freeman were trying to achieve the same goal for their large families. These two goals, accumulating land for a large family and keeping sons nearby, meant that the founding families of Ipswich used up their sizeable town grant more quickly than anticipated. There was still plenty of land farther west that could be acquired from Native Americans by intimidation or persuasion, but the lands of Ipswich were soon gobbled up by a rapidly growing population.

In addition to lands Edward Chapman acquired in subsequent town divisions, he held in trust for his children some land passed on by his first wife's father, Mark Symonds. Like many of his contemporaries, Edward was in no hurry to dispense the lands he held for his children. Perhaps it was a way to ensure continued filial piety, but the result was that beginning as early as the second generation, many young men found themselves settling on land to which their fathers still held title and thus technically lacking one of the requirements for freeholder status. It is perhaps no surprise then that as this second generation came of age in the 1660s, Ipswich and other communities gradually began to liberalize the strictures that had sharply divided people along economic and spiritual lines. Nonetheless, this gradual relaxation would not fully sate the desire of Edward's adult children and grandchildren to achieve independence. But the sons and daughters of the founding generation were for the most part reluctant to engage in a direct confrontation with their parents. Filial obedience was so prized in Puritan New England that it was technically a crime punishable by death for a minor child to disobey a parent, although there is no evidence this extreme punishment was ever carried out.

It certainly must have been scandalous then, when Edward's second son Nathaniel publicly challenged his father to relinquish title to his share of his grandfather Symonds's lands. The event that precipitated this challenge was Edward's decision to grant lands to his youngest son John, who was dying from smallpox, probably incurred during an epidemic that swept through New England during King Philip's War in 1676. Edward's intent, it seems, was to ensure an inheritance for his infant grandson, John Jr. The deed granted his son John these lands "during the term of his naturall life, and to his son, John Chapman, after him, if he liveth to the age of 21 years." Just two months later, John Sr. died, and the deed was held by his widow, Rebecca, until John Jr. came of age.[13]

Edward's decision appears to have prompted second son Nathaniel, who had married three years earlier but had not yet received land, to sue his father for control of the Symonds lands to which he was entitled. Why Nathaniel chose to do so is unclear. In all likelihood he and his young family were already living on and working family land to which he would eventually hold title. Perhaps he was impatient for the title of "freeholder" and the status it commanded; perhaps he feared his father would not distribute the holdings fairly. Whatever the reason, this challenge hurt his father deeply. From Edward's perspective, it was not as though he had failed to provide for Nathaniel's independence. He made sure that his second son had been educated with a skill, carpentry, and he certainly did not intend to deny Nathaniel his birthright. But like many of this first generation, Edward intended to wait until his death to distribute his lands. Nathaniel's decision to challenge his father was not breaking the law, but it was nonetheless a bold act of rebellion. None of Nathaniel's brothers joined him.

The founders of Ipswich had hoped to create a community of faithful Christians who would strive to be guided by the principles of love, charity, and obedience in all of their relations with each other. But they were realistic enough to understand that disputes would arise occasionally, and they established simple methods for resolving them. In this case, town authorities designated three impartial neighbors to hear Nathaniel's complaint and Edward's defense. They found in Nathaniel's favor and divided the Symonds inheritance into five equal parts, instructing that Nathaniel be allowed to choose "his share, according to his birth." Presumably, eldest son Simon would get to choose first. There is no surviving record of a deed transfer to Simon, but Edward Chapman's will, executed a few years later, suggests that Edward had "alreadye done for him beyond my other children."[14]

Edward died just one year after Nathaniel's rebellion, dispirited by the disappearance of family unity. In his 1678 will, he provided for his four surviving children. Simon received thirty pounds in addition to the lands he had already received, to be paid over five years, and also his share of the Symonds lands for which he had waited quietly and patiently. Nathaniel also received thirty pounds, but nothing more. Mary received thirty pounds and "one coverlet that is black and yellow." Samuel, his youngest surviving son, was designated executor and given "all my house and lands and chattels," with one condition: Edward's widow, Dorothy, would "have use of the parlour end of the house, both upper and lower roomes, with the little cellar that hath lock and key to it, with free liberty of the oven, and well of water, with ten good bearing fruit trees near that end of the house wch. She is to make use of, to have the fruit off them, also the garden plot fenct in below the orchard, and one quarter of the barne, at the farther end from the house, also to have the goeing of one cow in the pasture, and all during the time she doth remain my widow."[15] Edward certainly hoped that bonds of affection and filial piety would ensure that Dorothy would be cared for by her stepson Samuel, but nevertheless he felt the need to spell out her rights and possessions in his will. A well, a garden plot, the milk from one cow, and a small orchard would help provide for her comfort, even if it did not ensure her perfect independence.

It appears Nathaniel might have paid a price for his act of rebellion, for he received no additional lands beyond the Symonds lands he had already acquired. Edward died with little faith that his family would accept this distribution graciously. In a concluding paragraph, he wrote, "My will further is that all my children shall rest satisfied with what I have done for them, and if any of them shall through discontent, make trouble about this my will, that then they shall forfeit and loose what I have herein bequeathed to them or him, unto them that shall be so molested by them."[16] Edward's last wish for family harmony was, of course, one that would have been difficult to enforce through community law, but there is no surviving record of any legal challenges to his will.

This sad drama at the end of Edward Chapman's life is one of many similar stories that illustrate how difficult it was for New England's founding generation to realize their patriarchal vision for the colony even amid what seemed like an unlimited supply of land. It also provides a glimpse into what emerged as the perpetual struggle in colonial Massachusetts, and elsewhere, as landownership trumped a skilled trade as the preferred path to competency, economic independence, and freeholder status. Large family sizes, the practice of partible inheritance, and every New England son's desire for a freehold pushed the Puritan commonwealth deeper into Indian lands. Still, patriarchy, strong family ties, and the intensive farming practices of New England's farmers acted as countervailing forces slowing westward expansion. From the perspective of the Sagamore and the other Algonquin and Iroquoian peoples of the region, the transformation and expansion was startlingly rapid. In the wake of the defeat of Metacom's warriors in King Philip's War in 1676, Indian visibility as a separate people with rights to the land essentially ended in eastern Massachusetts.[17] Chief Sholan of the Nashaway people sold the deed to his people's lands in the wake of the defeat, and the English purchasers of that land renamed the place Leominster after a village in the west of England and soon began remaking them into a likeness of the English countryside. But as rapid as the transformation appeared from the Indian perspective, taking the broader view of English colonization of North America, settlers in the mid-Atlantic states pressed westward at a much faster pace, pushing up against, and even spilling over, the Appalachian spine by the middle of the eighteenth century. The fourth- and fifth-generation descendants of Edward Chapman would find it increasingly difficult to achieve the dual goals of acquiring a freehold and keeping close to family.

Edward Chapman's grandson, John Chapman, Jr., orphaned by King Philip's War, was among the first of Edward Chapman's heirs to leave Ipswich for lands further west. Although records indicate he remained in Ipswich to marry and start a family (the birth of his son John III in 1714 appears in the town's vital records), by the 1730s he appears to have sold his modest Ipswich inheritance and moved on to lands in the northeastern part of Billerica about twenty-five miles to the west. When the residents of that town sought to separate and form the town of Tewksbury, John Jr. was among the fifty founding families that the General Court granted nine thousand acres. In 1735, at the age of fifty-eight, and just four years before his death, he became one of the original covenanters of the Tewksbury church and a freeman in the new town. His son John III, made a reverse migration back to the place his great grandfather had first become a freeman, settling as a tenant farmer in the town of Topsfield, carved from the original Ipswich land grant. In 1760, John III died at the age of forty-six from smallpox contracted during a campaign against the Iroquois during the French and Indian War. Two years earlier, John III's eldest son, Perley, also succumbed to the disease. John III's younger son, Nathaniel, fourteen at the time of his father's death, and the only surviving male child, was taken in by an uncle and trained as a carpenter.[18]

LEOMINSTER

Nathaniel had a skill but no land. Finding work as a carpenter might keep him fed, but it did not qualify as true independence. While the prospects of earning a living as a tradesman had improved since the days of Edward Chapman, most young men saw trade work as a transitory means to a different end. Nathaniel Chapman aspired to being a land-owning farmer, for land was associated with independence and manhood.[19] When Nathaniel completed his apprenticeship, he headed west to the town of Leominster to find work as a carpenter, arriving sometime in the 1760s. In 1770, he married Elizabeth Simonds, a daughter of one of Leominster's first families. Elizabeth and Nathaniel set up house and a farm on a small piece of Simonds's family land, not far from the banks of the Nashua River, below hillsides covered with apple trees. In 1772 their first child arrived, a daughter they called Elizabeth; two years later, on September 26, 1774, as apples ripened on the hillsides behind their home, Elizabeth gave birth to a son named John, the fourth in the line back to Edward to bear that name.[20]

Leominster shared some characteristics with Edward Chapman's Ipswich. At its center stood a meeting house, which served as a place of worship and of politics. The founders had followed the same practices of other New England towns in dividing the land, and Leominster society was dominated by a small group of families who had been among the first to arrive. Farms spread out in all directions from the town center and were connected to the town by an array of rambling dirt roads, shaped by the landscape rather than imposed with an undue concern for symmetry.

But the Leominster of 1760 was also different from the Ipswich of a century earlier. While the Puritan church, now called the Congregational church, still dominated political and religious life in the community, religious conformity could no longer be imposed by town leaders. It was still true that church membership was the path to influence and standing, but Massachusetts had ceased to be a commonwealth shaped by a uniform faith. Leominster had its share of Presbyterians, Baptists, and the occasional Quaker, as well as many people who saw little use in attending church at all. Even in the established Congregational church, new ideas about faith were sowing division across the colony, as ordained ministers embraced or resisted the new emotionalism of the first Great

Awakening. Conservatives who balked at a movement that placed heart over head came to be known as Old Lights, while those who embraced this emotional awakening were called New Lights. Leominster's congregation sided with the latter and ousted a minister and his followers for views they insisted "did not hold or believe the essential divinity of Christ as it is revealed in the divine Word." The ousted minister defended his reasoned approach to the scriptures, asking, "Am I guilty of a crime? I am willing to be classed with Newton, and Milton, and Locke, and other good and great men, in the opinions which I hold. No one need be ashamed in their company."[21] Few Leominsterians welcomed this emerging diversity in religious perspectives, but they had come to tolerate it grudgingly.

Eighteenth-century Leominster differed from seventeenth-century Ipswich in other ways as well. While the region's farming families still focused much of their energy on providing their own needs, they were increasingly involved in production for local markets. New England families with eight or more children were not uncommon, which meant that land was scarce and increasingly valuable. Leominster's farmers had learned to use their varied soils and terrain to their maximum potential. Often this meant that they only grew what their land was most suited to and traded surpluses locally for other items their neighbors could produce more efficiently. Those with wet, low-lying lands reaped grasses to provide fodder for livestock; those with valuable upland soils tilled and manured to ensure maximum yields and to preserve their fertility. Apple growing was relegated to lands most difficult to exploit, the rocky clay soils of Leominster's hillsides. This emphasis on efficiency also accelerated the transition from beer to cider drinking. Farmers were reluctant to devote any portion of their scarce upland till soils to "growing" fermented beverages. By consuming cider instead of beer they could reserve their best fields for growing food, not drink. The result was a dramatic increase in cider production in the early eighteenth century. One village of forty families produced almost three thousand barrels of cider in a year; another village of two hundred families produced almost ten thousand barrels, or roughly fifty to seventy-five barrels per family annually. Eighteenth-century tax valuations often undercounted orchards, even as cider production was increasing. This was in part because of the "hidden" nature of New England's orchards. Tax assessors often failed to count apple trees randomly scattered throughout a pasture or small aggregations of trees clustered on a rocky hillside. Yet these unnoticed, diffused, decentralized orchards provided large quantities of fruit each fall for the cider mills. Most of this cider was consumed locally, but surpluses traveled farther afield. New England began exporting substantial quantities of cider to the West Indies in the late years of the eighteenth century.[22] These cider apples were for the most part seedling stock, typically small, crabbed, and bitter. A few generations later, Henry David Thoreau tells us, these orchards of "wild apples" fell victim to the temperance man's axe. But Nathaniel Chapman's Leominster abounded in these orchards, and they may have been the source of the first wild apple his son John ever tasted.[23]

Young John was born into troubled times. Tension between the colony's politically savvy freemen and the Parliament in London had been increasing since 1764. At times it had exploded into violent conflict. By the fall of 1774, the leaders of this emerging rebellion feared it would soon erupt into war. On September 21, five days before John's birth, a convention held in Worcester called on every New England town to reorganize its militia, to enlist one-third of all men between the ages of sixteen and sixty to serve in new militias, and to train them "to be ready to act at a minute's warning."[24] The original vision of the New England town militia called for an organization composed of property-holding citizens, but by custom the property-holding distinction had rarely been enforced. Militia service was not a duty all aging landowners were eager to reserve for themselves, and town militias by 1770 were increasingly made up of younger men, many of whom were not landowners. The order issued by the Worcester convention appeared to abandon this distinction altogether. Nathaniel Chapman signed up for Leominster's new "minutemen" militia as a private; any higher rank would have likely been reserved for a property holder. Even as the men of Leominster were stirred to action to defend their rights, first as Englishmen and then as "free men," the old traditions of rank and deference did not quickly melt away.[25]

When the alarm came early in the morning of April 19, 1775, that British regulars had left Boston and were marching westward into the countryside to find and seize stockpiles of arms, Leominster's militia answered the call. Nathaniel Chapman and his company headed east toward Concord, arriving too late to participate in the fighting that day, but they continued on to Cambridge where they joined with other town militias and began a siege of Boston. Enthusiasm for the cause of liberty ran high in 1775. Like other towns, Leominster initially sent off militia units that included citizens of higher social and economic standing, as well as those who owned little; middle-aged patriarchs with growing families marched alongside the young and unmarried. But as the conflict dragged on, it became a struggle fought primarily by the young and the poor. When the Leominster minutemen returned from Cambridge after eleven days service, Nathaniel almost immediately enlisted as a private with Asa Whitcomb's 23rd Foot Regiment, and he participated in the Battle of Bunker Hill two months later. Aged twenty-nine when he first went off to war, Nathaniel was among the eldest privates in his company. Perhaps he hoped that by proving his mettle, he might find new social standing in his town. And just one week after the Battle of Bunker Hill, Nathaniel and Elizabeth were admitted to full membership in Leominster's First Congregational Church. The timing was certainly not coincidental.[26]

In the spring of 1776, after spending the winter months with Elizabeth and his two young children, Nathaniel enlisted in Captain Pollard's company of carpenters, who joined with Washington's army in New York and helped construct the defenses required to repel a British invasion. Nathaniel could have stayed behind, and he had plenty of reason to do so. Elizabeth was pregnant with her third child, which might have been cause for celebration were she not sick with consumption (tuberculosis). While most New England towns saw a majority of their fighting age men go off to war, many served only one season before returning to their farms and businesses. The percentage of citizens who served extended duty, or for the duration of the conflict, was quite low. We can only speculate on the factors that contributed to his decision to leave his ill, pregnant wife and return to service—political convictions, a sense of duty to his comrades, and perhaps a hope that his continued service would help him become a landowner and increase his standing in his community. He must have made this decision with a heavy heart.[27]

By June, Elizabeth's condition had worsened, and she had not heard from Nathaniel since he had marched away with Captain Pollard's regiment. It is not hard to imagine the toll these uncertain days took on Elizabeth. There were daughter Elizabeth, six, and son John, not yet two, to worry about. The house was likely busy with family and friends attending to her in her sickness, keeping the household, and caring for the children. She turned to her faith to help her cope. Then, in quick succession, on May 31 and on June 2, she received letters from Nathaniel. Those letters do not survive, but the letter she wrote in response has been preserved, and it reveals a woman who is preparing for death and is aware that these words might be the last she will ever exchange with her husband.

June 3, 1776
Loving Husband,
    These lines come with my affectionate regards to you hoping they will find you in health, tho I still continue in a very weak and low condition. I am no better than

I was when you left me but rather worse, and I should be very glad if you could come and see me for I want to see you.

Our children are both well thro the Divine goodness.

I have received but two letters from you since you went away—neither knew where you was till last Friday I had one and Sabathday evening after, another, and I rejoice to hear that you are well and I pray you may thus continue and in God's due time be returned in safety. I send this letter by Mr. Mullins and I hope it will reach you and I should be glad if you would send me a letter back by him.

I have wrote that I should be glad you could come to see me if you could, but if you cannot, I desire you should make yourself easy as possible for I am under the care of a kind Providence who is able to do more for me than I can ask or think and I desire humbly to submit to His Holy Will with patience and resignation, patiently to bear what he shall see fit to lay upon me. My cough is somewhat abated, but I think I grow weaker. I desire your prayers for me that I may be prepared for the will of god that I may so improve my remainder of life that I may answer the great end for which I was made, that I might glorify God here and finally come to the enjoyment of Him in a world of glory, thro the merits of Jesus Christ.

Remember, I beseech you, that you are a mortall and that you must submit to death sooner or later and consider that we are always in danger of our spiritual enemy. Be, therefore, on your guard continually, and live in daily preparation for death—and so I must bid you farewell and if it should be so ordered that I should not see you again, I hope we shall both be as happy as to spend an eternity of happiness together in the coming world which is my desire and prayer.

So I conclude by subscribing myself, your

Ever loving and affectionate wife

Elizabeth Chapman[28]

By the time Nathaniel Chapman received this letter, events in New York were unfolding rapidly, and it is unlikely that he was granted leave to visit his ill wife. On June 26, Elizabeth gave birth to a boy, and she named him Nathaniel, but neither mother nor infant son were in good health. We cannot know if Nathaniel had received this news by July 1, when the British fleet arrived at Sandy Hook and Washington's army prepared furiously for the impending invasion of New York. We do not know how Elizabeth received the news one week later that the Continental Congress had declared Independence from Great Britain, or if she was conscious when reports of the British invasion of New York City on July 14 reached Leominster.[29] Elizabeth passed away on July 18, and the baby Nathaniel appears to have survived her only by a few weeks.[30] Young John and

Elizabeth Chapman were now motherless, and John was too young to carry any memory of her with him. Years later, Nathaniel shared their mother's final letter with his two eldest children, and for John, it would have to serve as a substitute for memory. His mother remained frozen in his mind as a model of perfect piety: her bond to his father was loving and ideal. Her final wish, that the marriage bond she had with Nathaniel would be restored in the next world, would have an important effect on John's later spiritual path.

SPRINGFIELD AND LONGMEADOW

For the next four years, the years when a child's first permanent memories are formed, John Chapman saw little of his father. Nathaniel probably returned to Leominster for the winter, after the long, discouraging New York campaign. The Declaration of Independence in early July had brought a momentary sense of optimism for the nation, but it was followed by the British rout of American forces at New York and the occupation of that city. With the deaths of his wife and an infant child he had never met, Nathaniel's personal sufferings went hand-in-hand with the setbacks of the new nation. New York City fell to British forces, and New Jersey soon followed. Around the same time, it appears that Nathaniel was forced to relinquish the lease on his small Leominster farm. The one bit of good news for the Americans was the surprise victories at Trenton and Princeton at the end of the year.

That winter was a winter of decision for Nathaniel Chapman. He had served the new nation and survived two seasons of war, at great personal cost. His Leominster neighbors certainly recognized his sacrifice, and few were likely to voice criticism if he had chosen to find local work as a carpenter and perhaps remarry and rebuild his family. But in the spring of 1777 he returned to service, with a promotion to captain of a company of carpenters and wheelwrights, to be stationed at the new arsenal being established at Springfield, Massachusetts, on the Connecticut River. The move to Springfield meant opportunity, but it also meant leaving Elizabeth and John behind with relatives. Family tradition suggests that they came under the care of their maternal grandparents, and it makes sense that they would come under the care of Elizabeth's family, which was spread about Leominster. Nathaniel's parents had passed away, and he had no other family in town. Young John Chapman probably spent his early years on the farms and in the orchards of Leominster.

When Nathaniel arrived there in the spring of 1777, Springfield was an unassuming village on the Connecticut River, about ninety miles west of Boston.

The arsenal there served as a storage and distribution point for arms, ammunition, clothing, and other supplies. Troops stationed there were responsible for procuring supplies, most significantly forage, in the region. The arsenal also manufactured cartridges and fuses, and it included an ordnance repair shop, where Captain Chapman would work for three and a half years. Springfield's location at the crossroads of a major east-west road, and along the navigable Connecticut River, made it an ideal location for distributing arms and ammunition throughout the Northeast. Far enough inland to be reasonably safe from British incursions, it was nonetheless well situated to provide support for military action at a time when much of the fighting was occurring in New England and New York. In 1777, when it opened, the arsenal was indeed near the center of action, as the British sought to "cut the head off the rebellion" by isolating New England from the rest of the rebellious colonies. In April, Benedict Arnold forced a British retreat from Ridgefield, Connecticut, 90 miles southwest of Springfield. In July, a serious threat 160 miles to the north emerged when Burgoyne's forces retook Fort Ticonderoga. One month later, American forces were able to slow down the progress of a British force with the help of Oneida Indian allies, 180 miles to the west at Oriskany. It was not until October with the American victory at Saratoga, just 100 miles northwest of Springfield, that the worst threats to the region subsided.

The Saratoga victory did not completely end the threat to the Northeast, but it did force the British to depend increasingly on local loyalists and Indian allies to carry on war in the Northeast. In the summer and fall of 1778, joint forces of American loyalists and Iroquois, under the leadership of Seneca war chiefs Old Smoke and Cornplanter, launched devastating attacks on communities in Pennsylvania's Wyoming Valley and New York's Cherry Valley, killing hundreds. General Washington, determined to strike back and permanently crush the Iroquois resistance, waited until the fall of 1779 to exact his revenge. He sent a force under Maj. Gen. John Sullivan from the Wyoming Valley into the heart of Seneca country in western New York, providing explicit instructions to carry out "the total destruction and devastation of their settlement." Seneca resistance melted in front of Sullivan's forces, who stopped at each abandoned Seneca village along the way and methodically destroyed large fields of ripening corn and, to the surprise of some soldiers, massive orchards of apple and peach trees also. The destruction of corn just before its ripening ensured a hungry winter for the Iroquois, but the extra efforts taken to chop down, burn, and girdle apple trees, which represented a decades-long commitment to these places, was intended to remove any incentive the Seneca might have had to return. "The Indians," Sullivan declared, "shall see that there is malice enough in our hearts to destroy everything that contributes to their support." The surprise that some of Sullivan's soldiers had expressed at the appearance of vast orchards of apples and peaches in the heart of Indian country was the result of white stereotypes of Native Americans, whom they regarded as primitive nomads and people incapable of adapting to new ways of living. In fact, the orchards they encountered were quite mature. Many Iroquois had adopted orchard cultivation decades earlier. But the memory of these Indian orchards proved to be short-lived. A few decades later, President Thomas Jefferson encouraged missionaries to go out and live among western tribes with pruning hooks and ploughshares, to "introduce" settled agriculture to them.[31]

The defeat of the Iroquois diminished the threat to the Northeast, and with the major locus of war shifting to the South, Springfield's armory became increasingly unimportant to the war effort. Springfield locals resented the costs of importing, housing, and feeding outside labor and saw a startling contrast between the operations of this new military machine and their own frugal Yankee habits. The arsenal's procurement activities exacerbated the problem. When it came to procuring bulky items like forage, the arsenal relied on the farmers from the nearest towns. Most were willing to contribute what they could to the cause, but eventually they found the demands for hay, peas, oats, corn, rye, and buckwheat too taxing. As the hardships of war continued year after year, even Springfield's most ardent patriots began to feel that they had contributed more than their fair share. They worried that the paper certificates that they were given in exchange for their goods might not hold their value, and they began to cast a critical eye on what they saw as the profligacy of the military bureaucracy in their midst. And in the fall of 1779, Springfielders petitioned the Continental Congress to demand an investigation into mismanagement of stores and the wasteful practices at the arsenal.[32]

In July 1780 the preliminary report from a congressional committee investigating the Springfield arsenal called for the excusal of "Ezekiel Cheever Esqr and Lieutenant Colonel David Mason," the senior civilian and highest-ranking officer at the arsenal. A line in this initial report suggesting that they were entitled to a year's pay and subsistence was crossed out, indicating there was some debate on the subject. The committee also authorized the Board of War and Ordnance "to remove any [other] unnecessary officers." In that same month the board concluded that the arsenal "has long been so very ill conducted that the benefits derived from it have been very inadequate." It recommended reforms, including that five officers and one civilian employed at the arsenal "be excused from farther service." Captain Nathaniel Chapman was one of the officers. The board appeared to take an agnostic stand on the charges of wrongdoing, advising that the men were "entitled to one year's pay and subsistence" unless Congress decided otherwise.[33]

Robert Price, author of *Johnny Appleseed, Man and Myth*, concluded from these reports (and the fact that there is no surviving record of Nathaniel receiving a land bounty for his military service) that Nathaniel Chapman was in effect dishonorably discharged, but the surviving evidence is more ambiguous. Certainly, Congress and the Board of War were displeased with the mismanagement and waste that plagued the arsenal. But by 1780, the arsenal at Springfield was not as critical as it had been earlier in the war. The decision to excuse six from service might have simply been a bureaucratic attempt by the board to take action that looked like reform, and circumstantial evidence suggests Nathaniel Chapman may have volunteered to leave. The same month the board issued its report, Nathaniel Chapman married Lucy Cooley, a local girl, and he settled with his new bride in the village of Longmeadow, just south of the Springfield line. It is easy to imagine that Nathaniel, recognizing a reduction in force was coming and eager to get back to civilian life and reunite with his two children, volunteered for excusal from service.[34]

Lucy Cooley, Nathaniel's eighteen-year-old bride, was a member of one of Longmeadow's leading families. The Cooleys were one of just four families who dominated Longmeadow; of the sixty original lots that extended along the main street of the village, about one-third were owned by persons with the last name of Cooley, and more were owned by persons related to the Cooleys by marriage. Like Leominster and other New England towns founded during the third generation, family connections now superseded religious ideology as the glue that held communities together; the numbers of last names present in these slightly younger towns was a fraction of what it was in older communities like Ipswich. Nathaniel's acceptance into one of Longmeadow's leading families is perhaps a further piece of evidence that his activities at the armory had not been obnoxious; by his bearing and demeanor, he must have overcome the initial prejudice he faced as one of the many strangers who worked there.[35]

John Chapman's new stepmother, Lucy, was barely beyond childhood herself. But she could certainly empathize with the losses her new stepchildren had experienced. Death was a frequent visitor to the Cooley household. Lucy was the ninth child of thirteen born to Mabel Hancock Cooley and George Colton Cooley. Five of her siblings had died in infancy. Lucy's mother Mabel named the last of these infants Submit, a signal that she was prepared to leave the child's fate in God's hands. Submit passed away shortly after delivery. But the recent loss of her father George was the most fresh in her mind. When Lucy was just sixteen, George had died following a smallpox inoculation. War had always brought smallpox. Nathaniel's father and older brother had succumbed to the disease during the French and Indian War, as had his great-grandfather a century before, during King Philip's War. The American Revolution ushered in a smallpox epidemic on an even larger scale, and in an effort to protect himself from this scourge, George agreed to be inoculated with a small dose of the live virus. While the practice of inoculation saved many lives, it could also kill. George took that gamble and lost. The common experience of losing a father during adolescence may have been one of things that drew Nathaniel and Lucy together; perhaps it also made her an empathetic stepmother to her new young charges.[36] After her father's death, her Uncle Jabez and Aunt Abigail Cooley remained especially concerned for their niece Lucy for the rest of their lives. Jabez and George had married sisters Abigail and Mabel. Lucy was not just the daughter of his deceased brother, she was also the daughter of his wife's sister. Jabez and Abigail were Lucy's uncle and aunt twice over, and they and their children felt a special responsibility for her for the rest of her life.[37]

Lucy had inherited a modest piece of her father's property, which fell under Nathaniel's control upon marriage. For Nathaniel this new marriage brought with it the hopes that he could realize his ancestors' dream of freeholder status and economic independence. Nathaniel and Lucy's property sat a few blocks off the main street of Longmeadow and less than a mile from the meeting house and town green. It contained a modest house and thirty-four acres of land—seven acres of tillage, three acres of meadow, ten acres of woodland, and an additional fourteen acres deemed "unimproveable." The Chapmans acquired a few livestock.[38] There was no orchard worthy of recognition on a tax assessment list, but the farm probably had apple trees nonetheless—a few in the dooryard and others scattered in the meadow or other parts of the land with limited utility. Nathaniel most likely worked as a carpenter to supplement the modest income of the farm.

Young John and Elizabeth joined their father and new stepmother soon after their marriage, and in December 1781 they welcomed a new brother, Nathaniel, into the world. The Chapman family grew steadily from this point on, with Lucy giving birth to ten children in a span of twenty-two years.[39] John grew especially close to Nathaniel and maintained his close ties to his older sister Elizabeth throughout his life. But the modest Chapman home soon grew crowded, and John had plenty of reason to seek the outdoors. It was a short walk down to the banks of the Connecticut River, a watery highway that increasingly linked Longmeadow to a much wider world. Undoubtedly, John picked up his skills in woodcraft and canoe-building in these Longmeadow years, as well as his Yankee ingenuity and lifelong habits of thrift and frugality.

While the river just east of his home and the now rapidly growing city of Springfield a short walk to the north offered young John Chapman glimpses of a wider, changing world, Longmeadow clung stubbornly to old New England ways. Longmeadow life was not without tension and turmoil, but very little had changed geographically and culturally in the village in the sixty-seven years since it had achieved precinct status in 1713. In many ways it more closely resembled Edward Chapman's Ipswich than John's native Leominster. The town still remained centered around the "long meddowe" green that gave it its name. It was there on the green that the town's meeting house rested. Its citizens remained relatively close spatially, with most residences stretched out along the north-south main street, which extended down from Springfield. The town was dominated by just four families, with Lucy Cooley's family being the largest.[40]

The religious landscape of Longmeadow harked back to Edward Chapman's Ipswich as well. The citizens of Longmeadow had not entirely abandoned the vision of their Puritan forefathers, who believed that the ideal godly community was one in which all members were "saints" who worshiped together in one church and subscribed to one orthodoxy.[41] The Longmeadow meeting house—like nearly all New England meeting houses—served the dual role as a place of worship and place of politics.[42] Longmeadow had built its first meeting house—an unheated structure, a thirty-eight-foot square, with primitive wooden benches—and hired its first minister in 1716. The goals of the Longmeadow villagers were exactly the same as those of the Springfield meeting from which they had separated—to create a community of like-minded individuals who strove for consensus in worship and in governance. This unity in worship was reinforced by their selection of a minister. Stephen Williams was just twenty-three when he took to the pulpit in Longmeadow; he did not relinquish that pulpit until he died sixty-six years later. Williams was still preaching there in 1780 when the young John and Elizabeth Chapman came to Longmeadow to join their father and stepmother.

Stephen Williams was in many ways a "typical" New England pastor. Educated at Harvard, he came from a family that had produced many ministers. His theology was well within the mainstream Congregational orthodoxy—in fact, he can fairly be described as a bit more conservative and traditional than most of his peers. But in one particular way, Stephen Williams was quite atypical. When he was a young boy, French-allied Iroquois attacked his village in an incident known locally as the Deerfield Massacre. They killed his mother and three younger siblings—all deemed too weak for the march to Canada—and carried Stephen away with his father and the rest of the family. Two years later Stephen was redeemed, but his sister Eunice was not. She was raised among the Iroquois, took an Iroquois husband, and converted to Catholicism.[43]

The Williams family made it a lifelong mission to win Eunice back, but she seemed content in her new life. Her life as a "savage" and her conversion to Catholicism were a double blow to the family. In the 1730s and 1740s Eunice made several visits to Longmeadow with her new family, but Rev. Stephen Williams's hopes that she would return to live among "civilized" people and to "the true faith" were never realized. On her visits, Eunice and her family did not yield to the requests of villagers that she cast off her Indian blankets and don English clothes, and she insisted on sleeping and cooking outside, in an apple orchard behind the Williams home.[44]

On one of her visits to Longmeadow, Stephen organized a special service in her honor. Eunice and her family sat politely but uncomfortably through the service as Stephen and the members of the congregation fervently prayed that she would see the light and return to the faith of her childhood. Their prayers were not answered; she died among the Iroquois and was given last rites by a Catholic priest. Although Stephen Williams was unsuccessful in his efforts to "redeem" his sister Eunice, he made it one of his missions to introduce English civilization and the Protestant religion to other Native Americans. Throughout his years in Longmeadow, he regularly took Indian boys into his home and transformed them (at least for the duration of their stay) into likenesses of English Protestants. Perhaps young John Chapman's first encounter with a Native American was with one of the young men who stayed with Williams.

The Longmeadow of John's youth was a community that held firm to the belief that there was but one way to live and only one path to heaven. It was a community that saw no virtue in tolerating dissent and only danger in religious pluralism. It also held that rank was essential to maintaining good order and good government and that a society without rank would surely descend into chaos and anarchy. Hierarchy in social status, Longmeadowans believed, was part of God's plan, and the structure of their meeting house reinforced the idea that some people were better than others. Even before prosperity enabled the citizens of Longmeadow to replace the primitive benches with slightly more comfortable pews, the struggle to create an order to the meeting house that would visibly demonstrate rank and status was begun. A committee of nine Longmeadowans, all from the village's leading families, took up the task of "dignifying the meetinghouse"—essentially creating a seating chart that reflected the status of families and individuals in the community. The process of seating was not uncontested—the chart was revised frequently from the first year of the church's existence all the way through the American Revolution—but the idea that there *should* be a fixed order in the meeting house was unchallenged.[45]

The religious enthusiasm now called by some historians the first Great Awakening swept New England in the 1730s and 1740s and in some places brought an egalitarian spirit to worship. But Longmeadow remained entrenched in tradition. In fact, it was in the very decade of this enthusiasm that Longmeadow replaced its primitive benches with carefully ordered pews. At the center of the meeting house, elevated five or six inches above the rest, was the pew designated to seat Rev. Stephen Williams's family.[46] A committee report from the year 1759 lists the seating arrangement at that time. In "the Great Pew by the pulpit stairs" sat the patriarchs of the Cooley, Colton, and Stebbins families. Other members of the Cooley family were scattered throughout the house, their seating determined by factors such as their landholdings, their age, their marital status, and the size of their families.

Longmeadow's minister eventually gained a reputation as the fiercest defender of old orthodoxies. When his own child was moved by hearing "awakened" preacher Jonathan Edwards to go out in the yard and begin shouting warnings to his neighbors that they turn away from sin and repent, Stephen Williams could find nothing wrong in his son's message, but he was unsettled by its emotional and unrestrained expression. When another Hampshire county town attempted to hire a young minister who, it was said, had preached the Arminian idea that individuals had control over their spiritual destiny, Williams led the effort to keep him out of the valley. His conservatism also made him a skeptic of the Revolution. He did read the Declaration of Independence from the pulpit, but a note in his diary suggests that act was not done voluntarily.[47] A short time later, leading members of the church insisted that Williams hire an assisting minister. The assistant lasted only two months before the senior pastor drove him off. Williams no doubt understood the War for Independence as a threat to traditional Longmeadow ways, and indeed it was. Throughout the country, the Revolution ushered in an age where concepts of equality, tolerance, and pluralism, once viewed as dangerous to good order, were gradually embraced as virtues, and the old values of rank, consensus, and conformity were increasingly undermined.

John Chapman was probably too young to recall any of the sermons at the "awakened" First Congregational Church in Leominster, which he left at the age of six. But the Longmeadow meeting and Williams's sermons were no doubt firmly implanted in his memory. But young John had more concerns than sitting still through the lengthy services; he had to get to know a father he had rarely seen, and a new stepmother. He had to find a place in a new and a rapidly growing family. And all of this occurred against a backdrop of economic uncertainty in the post-Revolution years.

SHAYS'S REBELLION

The Franco-American victory at Yorktown in September 1781 brought peace, but not prosperity, to the new nation. The farmers of the Connecticut Valley had endured years of hardship and sacrifice for the nation, but they emerged from the conflict racked with debt, only to find that agricultural prices were collapsing. It was certainly a precarious life on the modest Chapman farm. The farmers of Longmeadow could not imagine that things could get worse, but they did. Hardship prevailed across the Commonwealth of Massachusetts, but especially in the Connecticut Valley, where hard money was scarce. Longmeadow's farmers continued to rely heavily on a system of barter to make do in a period of economic chaos, with debits and credits recorded in the merchant's ledger and in personal daybooks and accounts being settled at the end of the year.[48]

In the statehouse in Boston, the conservatives who had taken control of affairs were more concerned with the commonwealth's creditworthiness than with the individual struggles of her citizens. The state had issued a sizeable amount of paper debt during the war, for livestock and forage and for a host of other goods and services. Indeed, the residents of Longmeadow and other Springfield area communities had contributed much, in exchange for paper. The value of that paper had rapidly diminished in the war years, as citizens lost confidence in the creditworthiness of their government. The region's farmers, desperate for cash, sold these wartime IOUs to speculators at a fraction of their face value, suspecting that the struggling state was unlikely to honor these debts.[49]

In 1784 the state legislature made the decision to commit to paying off war debt in full, with interest, and as rapidly as possible. Fearing that higher excise and import duties might dampen trade, the legislature made the fateful decision to place 90 percent of the needed taxes on persons and property. Moreover, the government insisted all taxes be paid in scarce hard currency, rather than goods. Under this plan, Massachusetts farmers bore most of the burden, and eastern speculators who had acquired most of this deeply discounted paper reaped most of the rewards, of the state's attempt to pay off its debt. In preparation for the levying of these taxes, the legislature ordered every Massachusetts town to compile a list of polls and property. Every male above the age of sixteen was taxed, as was all productive land. This was unwelcome news to the citizens of Longmeadow and every agricultural village in the Connecticut Valley, and local citizens' committees sent petitions to the politicians in Boston, pleading for a more equitable plan. Their pleas, however, fell on deaf ears.[50]

Longmeadow's farmers already labored under taxes heavier than anything they had paid as British colonists. These new levies increased their tax burden by five or six times. The residents of Longmeadow and other western towns were not entirely forthcoming when the polltaker arrived at their farms. While they could not undercount the acres they owned, they could declare that most or all of their "unimproved lands" were also "unimproveable" and therefore not taxable. Perhaps an extra sheep or swine could be forgotten in the count. It would be very easy to underestimate the productivity of a cider orchard, and there is evidence that Longmeadowans did just that. According to the poll, all of the orchards in Longmeadow were capable of producing just 725 barrels of cider annually, which amounted to only about six barrels of cider per household. Lucy's mother Mabel Hancock reported that her orchard would yield just five barrels a year. The typical farm family consumed between ten and fifty barrels of cider annually. The Longmeadow count of 1784, therefore, appears to be suspiciously low in a region where apple trees were ubiquitous. It is unlikely that Longmeadow, or any New England village in this period, was a net importer of cider.[51] Whether Nathaniel Chapman was completely straightforward with the tax assessor or he engaged in a bit of creative undervaluing, the result was the same. Unable to come up with the hard currency to pay his tax bill, he was forced to sell his farm to a wealthy neighbor, Nathaniel Ely, Jr., in 1785 for thirty dollars. It appears that Ely allowed the Chapmans to remain on the land, but now as tenants rather than landowners.[52]

By 1786, farm foreclosures were sweeping the Connecticut Valley, and many farmers who were also veterans of the American Revolution found themselves victims of a new government that was beginning to seem as oppressive as the one they had recently thrown off. One complained, "[I] have been greatly abused, have been obliged to do more than my part in the war; been loaded with class rates, town rates, province rates, Continental rates and all rates . . . been pulled and hauled by sheriffs, constables and collectors, and had my cattle sold for less than they were worth . . . The great men are going to get all we have and

I think it is time for us to rise and put a stop to it, and have no more courts, nor sheriffs, nor collectors nor lawyers."⁵³

The oppressive and unequal taxation of 1785 and 1786 wreaked havoc on the lives of many war veterans just like Nathaniel Chapman, even as it made a handful of eastern speculators very wealthy. Longmeadow was not a town reluctant to take up arms against unjust and oppressive governments. The Longmeadow militia had answered the call and marched off to Lexington and Concord in April 1775. But this crisis left them in conflict. Could they take up arms against the government they had recently risked their lives to build?

When news reached the region that a Massachusetts court would convene at Springfield in September 1786 to rule on a backlog of foreclosures, aging militiamen from across the region descended on the town to stop the proceedings by force of arms. Over the next several months, armed farmers across the state marched on courthouses in an effort to halt the foreclosures. In January 1787, a rebel army under the control of revolutionary war veteran Daniel Shays marched on the Springfield arsenal where Nathaniel had served, just a few miles up the road from the Chapman home. The attempt to seize the armory failed, and over the next six months a military force made up mostly of eastern militia pursued the rebels and put down the rebellion. Many of those who participated in the rebellion quietly slipped back to their farms, and a complete list of participants is impossible to reconstruct.

While the men of West Springfield, just across the Connecticut River from Longmeadow, rose up in mass to join the rebellion, Longmeadow's men were divided. Gideon Burt and Nathaniel Ely, Jr. (who had been able to increase his landholdings as a result of the new taxes) lined up with government forces to defend the Springfield armory. Alpheus Colton and John Bliss, both members of important Longmeadow families, joined the rebel march on the facility. There is no surviving evidence that reveals to us Nathaniel Chapman's choice, though he certainly shared the grievance of the rebels. But lining up with the rebels against his new landlord, Ely, would have been a very risky move.

The arsenal, of course, made sense as a target because of its stockpile of weapons. But symbolically, it also served as a fitting target for western farmers' rage. It had been, since its inception, a symbol of wastefulness and corruption, a hive buzzing with outsiders who failed to appreciate and understand the old ways of the surrounding villagers. Nathaniel himself had once been one of those outsiders and a target of local suspicion and resentment. But in 1786 he was another poor farmer who had served his country nobly during its fight for independence and who had been rewarded with nothing but hardship. Whether he actively supported the rebellion, stood to the side, or viewed it as an act of treason, he must have understood the rage that motivated the rebels.

In 1786 the Chapman family resided on the edge of a cauldron that was boiling over. Events at Springfield set off alarm bells throughout all of the colonies, where similar tensions were emerging between eastern elites and backcountry farmers. Young John Chapman turned twelve that year—too young to take up arms with either the rebels or the forces of the state that crushed them but old enough, perhaps, to understand the deep shame his father felt when he was forced to sell his land to pay his taxes. Not able to hold onto even a modest-sized farm, Nathaniel would never be the providing patriarch Edward Chapman had been. Twelve-year-old John Chapman surely understood that he would have to find his own competency in this world and, in all likelihood, move away to achieve it.

The turmoil of the revolutionary years accelerated changes already under way in the relationship between fathers and sons. The ideal of patriarchy—that fathers should provide their sons with a competency and that sons should remain deferential toward, and dependent on, fathers into their early adult years—persisted. But in reality, fewer and fewer fathers possessed the resources to realize that idea. As a tenant farmer, Nathaniel Chapman's only hope of providing his son John with a competency was to bind him into an apprenticeship where he might learn a skill. There is no surviving evidence that Nathaniel did, except for the fact that in 1790, the year John turned sixteen, he was no longer residing in Nathaniel Chapman's household.[54] There is certainly family tradition and some economic logic behind this path. Nathaniel himself, orphaned at a young age, was sent by his uncle to receive training as a carpenter. And with five younger children to feed in a very modest home, binding out his eldest son would have taken some pressure off the family finances. Apprenticeship contracts generally provided a cash payment to the parents, as well as some provision to provide for the apprentice at the end of the term.[55] An alternative possibility is that he was sent back to live with and work for relatives in Leominster, as his sister Elizabeth, who turned eighteen in 1790, was also absent from the Longmeadow household in 1790 and married a man from the Leominster area a few years later. In either case, it appears that John ceased to be dependent on his father at a young age, living elsewhere and earning his own bread. As a result, John Chapman is invisible in the historical record during his late adolescent years. While the tumultuous 1780s were certainly formative years in shaping young John Chapman's views on society, economy, and government, they were also the decade of his formal education. The scraps of neat handwriting he left behind in later years and his voracious hunger for reading material suggest that between household and farm duties, he found some time for formal schooling.

It appears that John Chapman was pushed out of the Connecticut Valley as much as he was pulled by the promise of the West. By the end of the eighteenth century, the valley offered few opportunities for the eldest son of a poor family. Land was scarce and expensive, and John could not expect to inherit a farm from his landless father. The ties that bound one generation of New England farmers to the next were rapidly unraveling. In the burgeoning cities, and in the lands opened to American settlement in the West after the Revolution—across the Allegheny Mountains, and over the Ohio River—new values ushered in by the Revolution and new economic realities were transforming the nation. John Chapman carried with him many of the values of his forefathers; chief among them, perhaps, was a desire for land and the independence associated with landownership. By necessity rather than choice, John was loosed from a crowded landscape and from many of the constraints of tradition and patriarchy. More by accident than intent, in walking west John and other young men like him played a small part in the reinvention of the American nation.

Each step took John away from the obsessions with family, rank, and status of the closed New England town; away from the whispered disapproval of a community scandalized by Eunice William's determination to dress in savage clothing, to cook under open sky, and to sleep under the stars; and away from the stifling religious orthodoxy of the Longmeadow congregation. Each step took him toward a world where equality of white men, at least, was assumed and worth was measured by what you might accomplish in a day with your hands and your back; toward a world where how you dressed, where you ate, and where you slept was nobody's concern; and toward a world that was filled with varieties of faith, rituals of worship, and ideas about God and heaven that the young John Chapman could have never imagined existed before he left Longmeadow. John Chapman walked west with hope and trepidation.

CHAPTER TWO

Becoming Johnny Appleseed

On March 29, 1853, members of the Warren, Pennsylvania, Lyceum gathered to hear Judge Lansing Wetmore, one of the town's most prominent citizens, deliver the first in a series of addresses on the early history of the county.[1] In his talk, the judge recounted the story of the arrival of one of the county's first white residents, a "tall stalwart Yankee" named John Chapman. According to Judge Wetmore, Chapman was residing in the Wyoming Valley on the eastern side of the state when he set out one November in the mid-1790s across the Alleghenies. He was outfitted like the typical frontiersman. A rope belt secured a rough pair of breeches around his thin waist. A hunting shirt hung loosely over this frame, and a folded wool blanket—which served as both a coat and a bed—rested on his shoulders, the corners pinned together over his chest. A rifle hung over one shoulder, and a tomahawk was secured to the rope belt around his waist. His feet were bare. And he carried a sack full of apple seeds.[2]

As this young man, about the age of twenty-one, climbed toward the summit of the Alleghenies, snow began to fall, and soon it was coming so thick and fast that he was obliged to seek shelter and build a warming fire. For three days the snow fell, and Chapman could not move ahead. When it finally stopped, he was forced to confront the direness of his situation, and perhaps to curse himself for his poor preparation for this journey. He was almost entirely out of provisions, the snow covered the ground three feet deep, and there were those bare feet. Whether he chose to push forward or turn back, he was a hundred miles from the nearest settlement in either direction. He cut strips of cloth from his wool blanket and fashioned primitive moccasins, but still his feet sank deep into the drifts with each step, and he could make little progress.

What he did next, Judge Wetmore hinted, was a testament to this young man's Yankee tenacity and ingenuity. He gathered as many small beech saplings as he could find, cut them down, and heated them over his fire until they were pliable, then he began twisting and tying them into a pair of makeshift snow shoes. The clumsy devices did the trick, and John Chapman pushed westward, driven by hunger and cold, descending from the mountains to the high Allegheny Plateau, arriving in Warren in early December. In the spring of the following year, after the snows had melted, John Chapman traveled a few miles west of Warren up the Brokenstraw Creek and planted his first apple tree nursery. The seedling trees provided stock for some of the area's first orchards, including those laid out by another early settler named David Mead. But the remnants of that nursery had been washed away long ago by one of the frequent freshets that gather their force from the melting snow on the region's steep ridges and flood the bottomlands along the Brokenstraw and other creeks in the spring. And that is how, Judge Wetmore told his attentive audience, the place called apple-tree bar got its name.[3]

When Judge Wetmore told of John Chapman's Allegheny crossing at the Warren Lyceum, perhaps a few members of his audience had heard a story about John Chapman from a grandparent or an old uncle, but for most the story was likely a novelty. The town of Warren was slow to develop, and few of the community's emerging middle class—the kinds of people who attended lyceums—could trace their roots back to the earliest days of settlement. Neither Judge Wetmore nor his audience, it appears, knew what happened to this peculiar young man after his brief sojourn in their neighborhood. Judge Wetmore said that Chapman had moved to Indiana and speculated that he had gotten involved in Indiana state politics. The moniker "Johnny Appleseed" was alien to them.

The story of John Chapman had undoubtedly made it to Wetmore's ears via some of the county's surviving first residents or the descendants of David Mead, who had purchased trees from Chapman. And there are a few facts in the judge's story that are contradicted by other evidence. John Chapman was not so tall, but of average height, perhaps five feet nine inches. Judge Wetmore dated his crossing of the Alleghenies in the late fall of 1797, though other records place him in the region at least fifteen months earlier.[4] But the story is of immense value because it is essentially the story of the apple tree planter's origins, and it appears to be an authentically local account, unaltered by the cross-fertilization that takes place when a folk legend gains national popularity. By 1853, a few stories of "Johnny Appleseed" had appeared in Ohio newspapers and journals and in one state history, but the legend had not yet reached a national audience. Wetmore's ignorance of Chapman's nickname and of his fate after leaving Pennsylvania affirm the story's local origin. As such, the Wetmore account, and a few other surviving western Pennsylvania traditions serve as the urtexts of the Johnny Appleseed legend.

It is worth setting Wetmore's story in its historical and geographical context, to provide a fuller picture of its significance and meaning. The starting point for Chapman's journey in Wetmore's account was the Wyoming Valley of eastern Pennsylvania. For more than forty years the lush farmlands of the Wyoming Valley had been the site of a fierce contest between Connecticut and Pennsylvania, both laying claim to the region based on original royal charters. Connecticut "Yankees" and Pennsylvania "Pennamites," each holding paper claims to the land from their home governments, waged an often bloody war for control of the region. The contest was not resolved in the courts until early in the nineteenth century.[5] Perhaps the young John, just approaching his majority, was drawn to the region by reading promotional material in the eastern press, which promised easy acquisition of prime farmland. In an effort to secure Connecticut's claim to the region "on the ground" rather than in the courts, promoters of the Connecticut-based Susquehanna Company had long been encouraging Yankee emigration and offering easy terms on land.[6] The Chapman family would certainly have heard these reports, as their Longmeadow home sat near the Connecticut River just a few miles north of the boundary with that state.

The Wyoming Valley during these years was a land abundant in orchards, with one observer noting that "every farm in the Wyoming Valley had an orchard. It was generally planted in the first cleared field, perhaps because other crops could be raised in the same field while the trees were growing."[7] The proliferation of orchards in the Wyoming Valley may have been accelerated by the Yankee-Pennamite contest. In a region where legal title to virtually every piece of land was contested, an orchard marked one's claim, and a mature orchard could provide evidence of the longevity of that claim. As a result, two- or three-year-old seedling trees could demand a price of six pence apiece in the Wyoming Valley, as new settlers sought to buy time toward a mature orchard.[8] If Chapman was indeed in the Wyoming Valley in the early 1790s, it is likely that this is where the seed of his plan to sell apple trees in frontier settlements farther west was planted.

In the mid-1790s there were several Indian paths that a traveler could take across the Alleghenies from the Wyoming Valley. One popular southerly route led to Fort Franklin at the point where French Creek joined the Allegheny River. If John first arrived in Warren, as Judge Wetmore's account suggests, he took a northerly route, which would have taken him along the path trod smooth by Sullivan's orchard-destroying army (some fifteen years earlier) for the first part

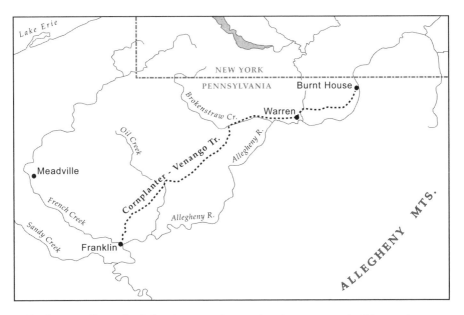

Northwestern Pennsylvania frontier, 1795–1804. Claudia Walters, University of Michigan–Dearborn.

of his journey.⁹ As he walked barefoot past the detritus left behind by Sullivan's force on a path made wider by its passing, did the young John Chapman, carrying the seeds for new orchards for the West, know the story of the Iroquois orchards and their fate?

The elaborate network of Indian paths that crisscrossed Pennsylvania were well engineered and thoughtfully laid out. Indian trailblazers cut paths that were relatively level and direct and as dry as possible, but they also sought to ensure the paths passed springs for fresh water at regular intervals. John was no trailblazer hacking his way through untouched virgin forest. He would not have had to take time each day building makeshift shelters for the night; every ten to twelve miles he could expect to find one built by Indian travelers. Unless John had managed to lose the path in the heavy snow, he likely would not have had to build his own shelter for protection, as Wetmore's account retells.¹⁰ But the American frontier folktale tradition is consistent in erasing Indian improvements from the landscape, be they orchards or carefully designed trails, shelters, and village clearings.

It is also worth considering his attire and accoutrements, which were—according to Wetmore's account—those typical of a frontiersman. A tomahawk and rifle were essential tools for such a journey, necessary for getting food and making shelter. The young John Chapman portrayed in the judge's story appears typical of the young frontiersman. Even the bare feet were not uncommon. Shoes were expensive and not easy to acquire in the West. Many adults, and even more children, went barefoot during the warm months of the year, saving a valued pair of shoes for inclement weather and special occasions. Nevertheless, undertaking a journey across the western mountains without shoes—particularly when setting out in the late fall—seems an especially rash decision. Was the young Chapman's shoeless state a result of extreme poverty, a purposeful stoicism, or youthful naiveté about what he would encounter in his walk west? Wetmore's account seems to suggest the latter. An aversion to footwear is a consistent element of the Ohio Johnny Appleseed tradition, but records from surviving store ledgers indicate that he frequently purchased moccasins during his Pennsylvania years, on some occasions acquiring multiple pairs at the same time.[11] The Indian moccasin did not offer the degree of protection of the European shoe and wore through quickly, but it was much more easily acquired and a staple item at frontier dry goods stores. Perhaps young John resorted to moccasin-wearing only in the most extreme conditions, or perhaps he continued to acquire pairs because they were an in-demand item he could easily carry with him and barter when necessary.

Finally, the timing of his journey, setting out in mid to late November, should be considered. The ideal windows for western travel were relatively narrow. In the spring, it was best to wait until the ice broke up on the rivers, but frequent spring floods could still pose problems when fording creeks and streams. The summer brought torrents of mosquitoes and biting insects, making woods travel difficult. In fall, a traveler would seek to find that short window between the departure of mosquitoes and the onset of winter.[12] Even given these constraints, John's decision to depart in late November seemed a risky one. His delay may have been prompted by his desire to wait until the end of cider-making season, as cider mills were the source of the apple seeds he carried.

Behind every cider mill in the fall one can find a pile of discarded pomace, the pressed, seed-filled pulp that constituted the waste product of cider-making. This offal had little value, beyond as hog feed, but John understood that the seeds it contained would have more value in new settlements across the mountains. It is hard to imagine a cider mill owner turning down a request from a poor young man seeking to recover seeds from this waste pile. His method of collection was likely a sticky one: fishing through the discarded pomace behind the mill, pressing his hands down into the still slightly moist pulp and liberating the seeds, his only company the buzzing wasps drawn to the sweet and sour aromas of the fragrant, rotting pile. Gathering the seeds into a sack, he would not be too particular about separating seed from pulp, as a little pomace left attached would help keep the seeds moist and serve as a kind of fertilizer when they were eventually planted. The aroma of rotting fruit, we can imagine, embedded itself under his fingernails and traveled with him as he walked north and west along the banks of the Susquehanna River. John may have made this cross-mountain journey several times during his Pennsylvania years, each time carrying new sacks of pomace-covered apple seeds with him.

In Judge Wetmore's account, John Chapman planted his first of many nurseries in the region along Brokenstraw Creek in the spring after his first Allegheny crossing. If Indian orchards were present in northwestern Pennsylvania at that time, their presence has escaped the historical record. The Seneca and Munsee Delaware who resided in the region in the 1790s were refugees recovering from decades of war and may not have felt sufficiently settled in the region to reestablish orchards. But John was not even the first white settler to carry apples into the region. That distinction belonged to another frontier bachelor named Cornelius Van Horn. Van Horn hailed originally from New Jersey but also had spent some time in the contested Wyoming Valley before pushing westward. In 1788 Van Horn set out with a small party of men from the Wyoming Valley, taking a southerly route across the Alleghenies to Fort Franklin at the conjunction of the Allegheny River and French Creek. From there the group traveled up French Creek by canoe to a place the Indians called Cussewago, and there they spent their first night sleeping under a large cherry tree. In the ensuing days they began the hard work of establishing land claims by plowing up soil, planting corn or potatoes, and setting out improvements. On this trip or a subsequent visit, Van Horn brought along some apple seeds and planted a small nursery along the banks of French Creek. If these trees managed to survive, they might, in a few years, serve as seedling stock for an apple orchard. A few of his companions erected temporary shelters, but Van Horn found an abandoned Indian cabin on his claim and took up residence in it. The group made several sojourns into the region over the next few years, returning east over the mountains before winter.[13]

It was not simply the harsh winters of the region that slowed settlement progress. Northwestern Pennsylvania was still contested territory in the 1780s and 1790s. The Iroquois Confederacy, which had long claimed sovereignty over the region, had been crushed by Sullivan's and Brodhead's raids during the American Revolution and ceded authority over most of these lands in a series of treaties signed with the new American government during the 1780s. But a western confederacy of Ohio Valley Indians continued to resist white expansion west of the Alleghenies and maintained the upper hand in the early 1790s. Van Horn and the other early French Creek settlers were repeatedly forced to retreat to the safety of Fort Franklin in the face of Indian raids.[14]

But Van Horn was determined to secure his claim. In the spring of 1791, despite reports of the presence of hostile Indians in the region, Van Horn set out again from Fort Franklin with two other men, determined to plough up a little of their land and plant some corn before returning to the safety of the fort downstream. Left alone in his field while his companions went for lunch, Van Horn was seized by two warriors. After binding his arms and marching him further upstream, his captors tied him to a tree and left him, presumably to forage or hunt for food. Van Horn recalled that he had a dull toy knife in his pocket and, managing to sharpen it on a stone a little bit, he was able to cut himself free from the tree but not to free his wrists. With his hands still bound, he set off in the direction of his claim, fully aware that his captors might soon return, discover his absence, and pursue him. But on the way back he stumbled across the seedling apple tree nursery he had planted a few years before. The nursery was choked with weeds, which he feared might expose his valued seedlings to destruction by the brush fires that frequently raged in the late summer. So, despite his predicament—hands still bound, not knowing whether his captors were in pursuit—Van Horn fell to his knees and began weeding his seedling nursery. It was in this condition that his companions found him and freed him from his fetters, before all decided it was best to retreat to Fort Franklin.[15]

Van Horn recounted this fantastic story himself, in an oral history conducted decades later. The story has a mythical quality to it and fits within the genre of frontier tales that portray their heroes as cool and collected even under the gravest threats. Whether it reflects a perfectly true account of events, it certainly highlights the value men like Cornelius Van Horn imparted to the humble apple tree. In another year or two, these seedlings could be transplanted to make an orchard, helping him secure his claim to the land; a few years after that they would begin providing Van Horn, and perhaps his future family, with an abundant source of food, drink, flavoring, and animal feed. A loss of this fragile creekside nursery to fire could have set his dream of domestic agricultural independence back several years.

By the time John Chapman appeared in the region five years later, the Indian confederacy that had effectively forestalled white settlement of northwestern Pennsylvania and Ohio had been defeated. General "Mad Anthony" Wayne's decisive victory over Indian forces at Fallen Timbers in the fall of 1794 laid the groundwork for the Treaty of Greenville the following year and eventually signaled the end of Indian resistance in northwestern Pennsylvania. Nevertheless, the world John Chapman entered was still much more an Indian one than a white one, and one in which a contest between Indian and white ways of getting a living from the land was still being waged.

BURNT HOUSE: THE INDIAN'S WORLD

Judge Wetmore's lecture on the first years of Warren offered a pretty tidy narrative of the settlement of Warren County. John Chapman's wilderness ordeal ended with his arrival at the town of Warren, where he presumably found shelter and the comforts of civilization. A few months later, he set out to plant his first apple tree nursery, and within a few years he had seedling trees to sell to the recently arrived farm families who rapidly would transform the region into a mixed husbandry landscape. Nowhere in the judge's narrative do Native Americans appear. But the "town" of Warren, where Chapman allegedly arrived, was but a paper fiction in the 1790s. Surveyed in 1795 by Andrew Ellicott, Warren's "town lots" were offered up for sale in Eastern newspapers in 1798. A single log cabin, planted on the site of an abandoned Indian village, its solitary resident a Holland Land Company agent named Daniel McQuay, was the only structure Chapman would have encountered at the spot where Connewago Creek met the Allegheny River.[16] There were a few dozen whites scattered here and there along the creeks that drained into the Allegheny south of McQuay's cabin, and a few more arrived each spring.[17] But the entire county only contained 233 white persons when the first census was taken in 1800. Ten years later, the county's white population had still not reached one thousand.[18] Warren remained a land company outpost, with no meaningful settlement for many years, and its first residents found it very difficult to impose the mixed husbandry regime on the landscape.

What would eventually be called Warren County was still Indian country when Chapman arrived. If he followed the primary Indian path down from the mountains, he passed by the Seneca village of Burnt House, six miles upstream before coming to McQuay's outpost. At Burnt House resided the Seneca chief Cornplanter and about four hundred of his people, and their settlement was the only place in the area that might be called a town in the 1790s. The people of Burnt House were also struggling to adapt to a new world in the 1790s, one in which Iroquois power was radically diminished, the new American government was ascendant, and their freedom to work the land in traditional ways was increasingly restricted.[19]

Like most Seneca towns in western Pennsylvania and New York, Burnt House had been destroyed by the American army in 1779, in this case by Colonel Daniel Brodhead's men as they marched north from Pittsburgh to join Sullivan's forces in the heart of Iroquois country.[20] By 1791, Chief Cornplanter and his people permanently resettled at Burnt House and began rebuilding. Cornplanter was among the first Seneca chiefs to make peace with the new American government after the Revolution and among the most pliant when it came to surrendering land to the victors. During the war between the U.S. government and the Western Confederacy, the Cornplanter Seneca remained neutral. By the 1790s, Seneca settlement was confined to a few small reservations in western New York and this tract at Burnt House, which the government deeded privately to Cornplanter to reward him for his cooperation. Cornplanter was able to extract only one meaningful concession from the U.S. government, and that was that his people still be allowed to engage in their seasonal subsistence activities—hunting, fishing, gathering, sugaring—across the region as long as they were able.[21] The government agreed, perhaps because officials were confident that a mixed husbandry regime of grain fields, livestock, and orchards would quickly dispose of forests and the predators that resided in them, bringing to an end the days of "the chase."

The majority of the men of Burnt House already would have left for winter hunting grounds at the time Chapman descended from the Alleghenies, and the village would have been left in the hands of women, young children, and the elderly. Burnt House occupied a narrow river bottom on the western bank of the Allegheny, ridges rising behind it, limiting western egress to a steep narrow trail. On both the northern and southern edges of Cornplanter's grant, the ridges pressed up against the river, restricting passage from north and south and permitting access from the east side only by canoe across the broad Allegheny. This self-contained valley is today submerged below the waters of the Kinzua Reservoir, the story of its Indian inhabitants commemorated only in a few campgrounds named in honor of Cornplanter and his half brother, the prophet Handsome Lake. But the antediluvian geography of Kinzua offered the people of Burnt House some refuge from the white settlers who were beginning to scramble over the Alleghenies in the late 1790s. At the center of the village was Cornplanter's home, two large rooms connected by a roofed breezeway, which served as a meeting house for his people and a residence for his extended family. Stretched out along the river were not traditional longhouses but smaller log cabins, each housing something less than an extended family. About sixty acres of cornfields along the riverfront were surrounded with split-rail fence, to keep out the handful of hogs and other livestock the villagers had acquired. The residents of Burnt House also planted corn on a few islands in the middle of the river. None of the surviving descriptions of Burnt House mention orchards in the settlement. Perhaps the chaos and instability that Sullivan's raid had wreaked fifteen years earlier had precluded the Seneca from reestablishing them.[22]

Cornplanter had concluded that Seneca survival in the post-Revolution world required some adaptation, and in 1795 he had acquired a water-powered sawmill and installed it on a small creek that descended from the steep ridges behind Burnt House. This innovation did not sit well with all of his people, but the Seneca at Burnt House were able to earn some money selling lumber to the Holland Land Company and even to the U.S. Army base at Franklin.[23] Near the center of the village, a large wooden statue of Tarachiawagon, the good twin, the Skyholder, and the Iroquois divinity, towered above the settlement. During much of the year in the northern Allegheny plateau, the clouds press low against the ridges and river bottoms, and if Chapman gazed upon the village from the snow-covered hillsides on the eastern banks of the Alleghenies, it might have appeared that the Skyholder was doing just what his name implied.[24]

Centuries-old tradition and innovation jostled uneasily in Burnt House, as Cornplanter's people grappled with their diminished power. One scholar has referred to Burnt House as a "wilderness slum," suggesting its residents were contained and had completely surrendered to despair. But the people of Burnt House still outnumbered their white neighbors in the mid-1790s.[25] They continued to spread out across the region to participate in their seasonal subsistence activities. They engaged in trade with each other and their white neighbors. Burnt House and the surrounding region was still Indian country at mid-decade, and despite the press to adopt white ways of living, Indian ways still prevailed.

Whether John, hungry and cold, sought food and shelter at Burnt House that winter, and whether he was welcomed or rebuffed, we do not know. But as one of the earliest white settlers in a region still dominated by Indian peoples, his encounters and exchanges with the people of Burnt House were certainly frequent. The seasonal mobility that was critical to traditional Seneca mixed-subsistence lifestyles meant that the people of Burnt House were often spread out across the region. In the Cold and Very Cold Moons of winter, they dispersed in small groups to hunt; in late January or early February, they returned to the village to celebrate the Midwinter Moon Ceremonies and give thanks for the game the forest had yielded. During the Sugar Moon they once again set out in small family units to tap maple trees, then during the Fishing Moon returned to their riverside village to construct weirs for gathering fish. The Iroquois calendar reveals the variety of activities that comprised the Seneca year and kept the people of Burnt House in a cycle of collection and dispersal: the Planting Moon, the Strawberry Moon, the Green Bean Moon, the Green Corn Moon, the Harvest Moon, and the Hunting Moon.[26] Other activities were responses to events that did not happen on a predictable schedule. In the years when the forest yielded a heavy mast, villagers gathered nuts; when vast flocks of passenger pigeons descended on the region to nest, they eagerly awaited the arrival of the young, then knocked over the nests, gathering the abundant squab as they fell to the ground, after which they would feast on nothing but squab for weeks.[27]

While the young John Chapman may have met a few of the Indian youth that Rev. Stephen Williams had taken into his Longmeadow home to convert to white ways, the Cornplanter Seneca were probably the first unassimilated Indians he had ever encountered. In the legend that developed around him after his death, John Chapman was beloved by Indians, who immediately recognized him as a kind soul with pure intentions. But the Pennsylvania oral traditions that have survived suggest that not all of John Chapman's interactions with the Indians of the region were peaceful. One resident of the region recalled John Chapman telling tales of his "hair-breadth" escapes from Indians in his first years along the Allegheny.[28] And this should come as no surprise. John Chapman's relations with his Indian neighbors, like those of all the early white settlers, were undoubtedly complex. Northwestern Pennsylvania in the 1790s was still a "middle ground" where Indians and whites lived in close proximity, in mutual dependency, but also in tension. Whites changed Indians, and Indians changed whites. Indians adopted some white technologies, dress, and ways of getting a living, and whites like John Chapman did so in turn. Indian trails with Indian shelters had made his westward walk possible; the Indian snowshoe saved his life in the high Alleghenies; and the Indian canoe helped him navigate the rivers of the Allegheny watershed.[29]

Local environmental realities often determined the pace and direction of cultural adaptation, and the landscape of the northern Allegheny watershed proved to be more suited to Indian ways of living than to white ones. In the first years after whites began arriving in the region, and in some places for decades, it appeared that the white newcomers were doing much of the adapting. A ledger from a dry goods store on Brokenstraw Creek near Warren recording purchases for the years 1795 to 1800 reveals that both the white and Indian customers paid for goods mostly with pelts and furs. Few offered surplus grain for payment, and the region was never dominated by agricultural pursuits.[30] Despite the confidence of white settlers in the ability of their mixed husbandry regime to bring wealth, much of Pennsylvania's Allegheny plateau proved to be less than ideal for it. Forest-covered ridges with thin soils rose up not far from the banks of the Allegheny, and even the fertile bottomlands along the region's many waterways were vulnerable to spring freshets, which could destroy a crop shortly after it was planted. In the short term, John Chapman and other white settlers found better economic prospects in hunting. "The calls of nature for food and clothing were imperious, and the slow returns of agriculture were unequal to the urgency of those demands," explained one traveler who visited the region in the 1810s. "Necessity compelled them to neglect the labours of the field for the chace. Habits, subversive of industry were thus acquired; simply to *live* bounded the view of their ambition." From this traveler's perspective, the white residents of the region had descended to live like Indians. "Hunters never grow rich," he moralized; "he who depends on the forest for his meat from year to year ought not to be enrolled as a citizen."[31]

As long as white settlers remained hunters rather than husbandmen, the traditional subsistence methods of Cornplanter's people faced only a limited threat, but white settlers also found profits in timbering, and the plateau's many navigable waterways, which descended to Pittsburgh and ultimately all the way to New Orleans, made the region's forests ripe for exploitation. Newcomers with dubious claims to the land, or no claim at all, clear-cut parcels and then moved on to new timber before anyone could challenge their right to it. John Chapman may have played some role in the denuding of the region's forest, as one local legend claimed that he could cut more trees in a day than most men could in two.[32] Before he became Johnny Appleseed, he might have been on his way to becoming Paul Bunyan. Timbering proved a serious threat to the people of Burnt House. Forests provided a home for game and a source for nuts and wild berries. Soil from cleared fields ran into the region's streams, exacerbating flooding, diminishing water quality, and harming the fish supply.[33]

Regardless of his long-term intentions—perhaps finding a suitable farm and homestead site, as well as places to plant his apple seeds—in the short term John, like all white settlers, had to adopt the highly diversified subsistence strategies of his Indian neighbors. One oral tradition tells us that he subsisted an entire winter on gathered butternuts; in February 1797 he purchased a "speck gimblet," a carpenter's tool designed for boring holes in wood, from John Daniels's store on Brokenstraw Creek.[34] As this was just before the region's many sugar maples would begin to sap, it is possible he was imitating his Indian neighbors and tapping sugar trees. It is not hard to imagine that he might have found himself intruding upon Indian sugaring grounds and invoking their hostility.

Even as John Chapman and his white neighbors were forced to employ a mixed subsistence strategy to get a living from this land, Cornplanter and some of his people were receiving an education in white ways of getting a living. In 1798, at the invitation of Cornplanter, three Quaker missionaries from Philadelphia arrived at Burnt House to teach the Seneca how to farm like white people. The Quakers were more interested in transforming Seneca lifestyles than indoctrinating them into their religious worldview. Turning the Seneca into an "industrious" people was their primary goal. In order to do this, they encouraged sobriety and discouraged the Seneca from wasting so much time on ritual "frolicks," which sometimes went on for days. Most significantly, they sought to reorder Seneca gender roles. Seneca men needed to abandon hunting and take the place of women in the fields; women could be taught various domestic arts that enabled them to contribute to the household income. The Quakers were convinced that these reforms were the path to Seneca prosperity.[35]

Ironically, no white farmers in the region were finding success following this path. While the missionaries were trying to persuade the Seneca to live like white people, John Chapman and many of the other whites in the vicinity of Burnt House were surviving by living like Seneca, in part because of geographic and economic necessities. An environmental transformation, which would be ushered in by deforestation, needed to occur before the lifestyle of the self-provisioning farmer was possible. Even then, given the limits of the land, the profits would fall only to a few.

American leaders had mixed reactions to the open lands of the West. For Thomas Jefferson, this open land ensured that future generations of Americans would be able to arrive at independent manhood by becoming masters of their own soil. In Jefferson's mind, the landowning yeoman farmer, imposing a mixed husbandry regime on the landscape, was the foundation of citizenship and the embodiment of republican values. But while Jefferson's vision was shared by many American leaders, a more negative view of the frontier existed alongside it. East Coast elite who traveled to the frontier often commented disapprovingly on the dissipated, savage state of its white residents and worried about the decivilizing effect of the wilderness. The frontier was to them a contest between civilized men and the forest—could the settlers transform the wilderness into an orderly, domestic landscape before it transformed them into savages? They saw the continued dependency of frontier people on hunting for survival as a major obstacle to advancement. It was the Indian's stubborn resistance to abandoning hunting that kept him from becoming civilized, and it was the need of white husbandmen to resort to hunting to sustain themselves that threatened to decivilize them. J. Hector St. John de Crèvecoeur, who praised the political and moral virtues of the American farmer, worried incessantly about his brothers on the frontier. "As long as we keep ourselves busy in tilling the earth, there is no fear of any of us becoming wild," Crèvecoeur asserted. "It is the chase and the food it procures that have this strange effect."[36]

One aspect of the Johnny Appleseed myth is the idea that Chapman rejected the hunt as a way to get a living. He never carried a rifle, the myth tells us, and he was loath to harm any living creature, which implies he was a vegetarian. But the Chapman in Wetmore's account did indeed carry a rifle, and committing to a vegetarian diet on the frontier would have been extraordinarily difficult. Although there are no surviving records that document Chapman paying for purchases with pelts and furs, as many of his neighbors did, he did purchase salted pork and gunpowder at the Holland Land Company store in Franklin.[37] The image of Johnny Appleseed as frontiersman who rejected hunting may have been rooted in part in historic American ambivalence to the activity, from Crèvecoeur's time down to our own. From its inception, the Johnny Appleseed story has served, in many ways, as a countermyth to the legends of hunters like Davy Crockett and Daniel Boone.

BROKENSTRAW: NAVIGATING CONTESTED TERRAIN

That the majority of the early white settlers of northwestern Pennsylvania had a difficult time transitioning to the landowning yeoman farmer ideal was in part the result of the region's limited suitability for mixed husbandry farming. The mobile, diversified, mixed subsistence strategies of Native Americans held more promise in the short term. But political barriers, as great or greater than environmental ones, also played a large role in the region's slow development. Even as the new settlers adapted to environmental realities by embracing hunting and other subsistence activities, they generally retained European ideas about private property and upheld landownership as their highest goal. The twenty-one-year-old John Chapman undoubtedly shared the view of his forefathers that landownership was the key to independent manhood, political citizenship, and domestic happiness. Among the most powerful myths of American history is that the seemingly unlimited open land on the frontier meant that landownership was a dream achievable by anyone willing to go west and claim it. In fact, John Chapman and his neighbors soon discovered that the law often confounded pioneers' attempts to become landowners. When John Chapman set out his first apple tree nursery along the banks of Brokenstraw Creek, he was almost certainly seeking to establish a land claim there, but that proved to be a more challenging task than he imagined.

John's presence along Brokenstraw Creek was mentioned in Judge Wetmore's story but also affirmed by repeated entries in the ledger of dry goods merchant John Daniels, who kept a small store near the creek in the last years of the eighteenth century.[38] The Brokenstraw flows east toward the Allegheny across a rare fertile bottomland in a region defined primarily by its steep stony ridges. Its clear waters run over a bed of smooth gray stone, and the stream braids and divides periodically as it flows toward the Allegheny, forming several small islands and gravel bars. In spring, when the snow melts, the Brokenstraw is easily navigable by canoe; by midsummer, a person traveling this method might have to step out periodically and drag his boat over shallow stretches. An angler today would immediately identify it as a good trout stream.

The creek earned its name from the tall grasses rising thickly along its banks—possibly prairie cordgrass, reed canary grass, or a mix of riverine species—which folded over or "broke" under the weight of the first heavy winter snows, forming dense mats of vegetation that made travel by foot or horse along its shores quite challenging.[39] This broken straw however, also choked out incursions of the forest, and when burned off, the flats—especially the broadest ones near its mouth on the Allegheny—easily could be converted to planting land. In a region defined largely by steep forested ridges, the flats along the Brokenstraw were valuable real estate. This "broken straw" was such a distinctive feature of the creek that it provided the name for the creek in several languages: for the Delaware, it was Paks-kalunska, for the Seneca, it was known as De-ga-syo-noh-dyah-goh. When a French expedition under Celeron de Bienville claimed the land for France in 1749, they dubbed it Le Paillee Coupee. Forebodingly, a *paille coupée* in European diplomacy was a symbol of discord or broken agreements. And in the last half of the eighteenth century, the Brokenstraw was the scene of endless conflict, as Indians, Frenchmen, Englishmen, and finally American speculators and squatters fought over ownership of the land.

That John Chapman was attracted to the lands along the Brokenstraw is no surprise. It was an ideal location for setting out his seedling trees and had some prospects for farming, but if he desired to acquire his own piece of ground, he would soon find that the path to landownership was not a straightforward one. The settlement of land in northwestern Pennsylvania was subject to the rules set out in the Pennsylvania Land Act of 1792, as confusing a piece of legislation as any ever created by democratic institutions. The law's authors attempted to serve two conflicting legislative interests at the same time. The first was to reward friends and political allies with an opportunity to amass vast wealth through land speculation; the second was a desire to see the Allegheny lands quickly occupied by actual settlers.[40] To meet these goals, the act provided two paths to land titles. The first allowed persons with the financial means to purchase warrants for 400-acre tracts from the state, have the land surveyed, then settle, or "cause to be settled," each tract within two years of purchase of the warrant. A settlement claim required some evidence of improvement; specifically, a primitive cabin had to be constructed and improvements made to at least two of every hundred acres claimed by warrant. Beyond the construction of a log home, evidence of improvement might include cleared forest, ploughed and planted ground, gardens cordoned off by split rail fences, and the establishment of orchards.[41] John Chapman's seedling trees, once they reached the age of two or three, would be valuable to any person trying to meet the settlement claim requirements. Within one week of the passage of the 1792 Land Act, members of Pennsylvania's political elite had purchased warrants on tens of thousands of acres of Allegheny lands. Within a year, land companies had acquired warrants on over a million acres in the region.[42] John Nicholson, a stakeholder in the Population Land Company used his position as comptroller-general for the commonwealth of Pennsylvania to claim by warrant virtually the entire Erie Triangle of lakeshore lands Pennsylvania had recently acquired from New York. Nicholson made sure that the few politicians in a position to challenge this action, including Governor Thomas Mifflin, would also benefit if they let his actions go unquestioned.[43] Also, General William Irvine, one of the Population Land Company's major stakeholders, had been commissioned to conduct the first scouting survey of the region, and he therefore knew precisely where the region's most valuable land lay. Irvine also offered his services to a group of Dutch investors incorporated as the Holland Land Company, which had already been busily gobbling up land in western New York, and he helped them acquire warrants on almost a half million acres in northwestern Pennsylvania.[44] Within a year of the Land Act's passage, companies or well-connected individuals like Irvine held warrants on virtually all of the lands considered valuable across the northern Allegheny watershed. During that first year, poor whites who wanted to make claims on these lands through settlement and improvement, were prohibited from doing so by Indian wars in the region. When the danger of Indian attack subsided several years later, it appeared that the majority of those interested in settling in the region would have to do so as clients of the land companies.

While the agents of the companies used the most questionable of tactics to acquire land, taking out warrants in the names of friends, relatives, and fictional individuals in order to hold them, their path to vast profits was not as simple as they believed it would be. The settlement provisions of the act meant that their warrants were not completely secure until they could demonstrate that they had improved these lands. William Irvine and the other American agents of the Holland Land Company advised the Dutch investors that this would not be a problem. The plan was to induce the poorest citizens—landless families like John Chapman's—to settle and improve the lands for them. The company put up additional money to help these potential settlers with relocation expenses; they provided them with tools and provisions needed and even granted them outright title to a portion of each tract once they had helped the company meet the settlement and improvement conditions of the act. They also gave these settlers the option of purchasing additional lands from the company, on easy credit terms, with no payments due in their first years.[45] Irvine and the land company barons expected settlement to proceed quickly, their titles to be affirmed and the profits from the transfer of titles to the actual settlers to result in huge profits within a few short years. But this scheme had an important catch. The land barons might have temporarily gained legal control of vast swaths of territory, but they were completely dependent upon the poor they hoped to profit from if they were ever to see a return. And the same Indian troubles that gave the land barons cover for carrying out their land grab eventually threatened to be their undoing. The western confederacy of Indians, which had launched these attacks to protect their Ohio Valley lands, proved more resilient than Eastern politicians anticipated. In northwestern Pennsylvania even the boldest and most determined settlers, men like Cornelius Van Horn, found themselves forced into repeated retreats to the protection of Fort Franklin.

In 1792, and for several years after, the land companies found few settlers willing to risk their lives to acquire land. By the time the Indian wars had ended and the worst violence had subsided, almost four years had passed since the purchase of warrants, and none of the land had met the two-year settlement and improvement provisions. The companies relied on an escape clause in the act to maintain their claims on millions of acres of unsettled and unimproved land. The law stipulated that if "enemies of the United States" had made meeting these conditions impossible, the two-year timetable for improvement would be void, or at least delayed.[46] The improvement clause now rested on a legal debate: were land company efforts to "cause" these lands to be settled during the Indian wars enough to validate their titles, or was the two-year clock on settlement and improvement simply reset once the threat subsided? And if the latter, when precisely had the enemy threat ended?

Over the mountains in the Wyoming Valley and to the south around Pittsburgh, lawyers who sniffed opportunity had begun to tell potential settlers that the warrants acquired by the companies in 1792 were no longer valid, and were they to move to the region, build a house, and set about improving the land, these lawyers would help them acquire title by right of improvement. Yankee farmers in the Wyoming Valley were especially susceptible to these entreaties. They had been fighting a decades-long contest on the ground and in the courts to secure their claims in the region. By the 1790s, the ground war was over, it had become a legal contest, and the commonwealth of Pennsylvania had the upper hand. Many claimants in the Wyoming Valley feared it was only a matter of time before courts dispossessed them of their long-held farms. The promise of acquiring new lands by improvement on the other side of the mountains was tempting. Rumors began circulating as early as 1795 that a tidal wave of grasping Yankees from the Wyoming Valley might spill over the mountains and begin making claims. By late 1796 the land barons began taking this possibility seriously.[47]

It is in this context that Chapman's appearance in the region in 1796 must be understood. Land company agents, and any others who had followed the warrant-then-settlement path to ownership would have been, initially at least, deeply suspicious of the intentions of a Yankee coming from the Wyoming Valley. One person who may have cast a suspicious eye on John Chapman in the spring or summer of 1797 was General William Irvine's son, Callender. Callender and John both found themselves on the Brokenstraw that year. Callender arrived in May 1797, and with the aid of the family's black servant, Tom, hastily erected a cabin and began setting out improvements on lands his father had claimed by warrant within three days of the passage of the 1792 Land Act. Despite being roughly the same age, John and Callender had quite different backgrounds. John was of Puritan Yankee stock that could be traced back to the great English migration of the 1630s; Callender's father, an Anglo-Irishman who served as a surgeon in the British Navy during the Seven Years' War, decided to stay in North America when that conflict ended. Both young men's fathers had served in the Revolutionary War, but the war brought very different personal results to each family. Nathaniel Chapman had risen to the rank of captain but lost two farms during the years between the signing of the Declaration of Independence and the ratification of the Constitution. In 1797 he was a tenant farmer with a large family to care for and concerns about how he might provide each of his sons a competency. William Irvine had acquired wealth and property by marrying into an established family in Carlisle, Pennsylvania, and had fostered important political connections during his service as a high-ranking officer during the Revolution. Those political connections paid off when he received the commission to lead the survey of northwestern Pennsylvania lands. When the legislature signed the law that opened these lands up for sale, William Irvine quickly snatched up several parcels at the mouth of the Brokenstraw, lands he knew from his travels in the region were the best available. Callender appeared to have all the advantages family money, and power could provide; John, rough-clad and barefoot, had no such resources.[48]

Callender's decision to go west that spring was partly motivated by a desire to prove himself to his overachieving father. Callender had been an indifferent student at Dickinson College back in Carlisle, almost dropping out before completing the two-year degree. And by 1797 he was not faring much better in a legal apprenticeship his father secured for him with Jared Ingersoll, one of Philadelphia's most politically powerful Republican lawyers. When his father informed him that he feared the lands they had claimed but not yet settled along Brokenstraw Creek were threatened by a potential influx of Yankee squatters, Callender seemed happy to have an excuse to abandon the study of law and to prove his usefulness in another way.[49]

During the summer of 1797, as Callender and Tom worked to build a primitive cabin and put in place some other improvements, father and son exchanged a series of letters about the situation on the Brokenstraw. Callender complained that the month of May had been so cold that for several nights in a row the water in his kettle, placed only a few feet from the fire, had frozen solid by morning. He wrote that there were rough-looking men crossing past his primitive homestead every day, and he feared they had intentions of stealing the family's land. The majority of the white men he had encountered were "not many removes from savages."[50] Given the small number of whites residing on the Brokenstraw in the summer of 1797, John Chapman was probably one of these persons upon whom Callender cast a suspicious eye.

William advised his son to get as many improvements done as he could and to return to the safety of Carlisle before winter, hoping that the presence of improvements would be enough to protect the land from squatters. But Callender had his doubts, writing that "it will be necessary in my opinion for someone to remain here all winter as ... many men look upon Brokenstraw with a wishful eye, and are determined to take possession should an opportunity offer." William was a protective father, as evidenced by his decision to send a servant along with Callender and his repeated letters of concern. He feared his son was endangered by not only the Wyoming Valley Yankees but also the late summer fogs that settled in the riverbed, which he considered a disease-causing miasma, cautioning Callender "to try to keep your hut dry & airy" and not to dabble near the creek after June. Father and son finally settled on a plan to contract the services of the one respectable neighbor, John Andrews, to look after the land in his absence, even offering Andrews the service of Tom in exchange.[51]

The Irvine letters provide a glimpse of the depth of suspicion, hostility, uncertainty, and populist assertiveness that would plague the Allegheny region for the next decade, as land barons and their agents squared off against ordinary citizens seeking to acquire a farm. Despite the family's continued concerns about their Brokenstraw lands, other opportunities meant that the Irvines would be absentee landowners for many years, as Callender accepted a series of military and political appointments that kept him from returning to the area.[52] In his absence, he relied on neighbors and hired hands to look after his lands. For a few years he employed a young man named Matt Young to occupy his land and continue with the improvements, but challenges to his ownership were constant. In June 1798 Young reported to Callender that a Charles Holeman appeared with a lawyer on Irvine's largely unimproved tract on the north side of the river, telling him that "in the name of the Commonwealth I forewarn you not to cut a stick on this Land muttering a great deal more about forseeable possession and threatened to make me pay smartly for working there." Years later, the son of an early Holland Land Company official recalled that northwestern Pennsylvania in these years was a place where "the assertion of popular rights was carried far towards the verge of ultraism ... and made almost every man in the community a student of law, and an imaginary, if not real professor, of the science of legislation."[53] In this instance, however, Mr. Holeman was convinced to back down once he was made aware of whom precisely he was challenging, and he quickly reconsidered his claim. The Irvines were one of the most powerful families in the state, and at that moment Callender was the captain of a militia unit station just forty miles downstream at Franklin. The challenge resolved, Callender instructed Young to put in a crop of potatoes and a small peach orchard, surrounded by a split rail fence, on the land the brash Holeman had tried to claim.[54]

The following year, Callender contracted with Young once again to reside on his Brokenstraw lands and make additional improvements, but it appears that Young did little more than consume Callender's provisions and request more supplies and money. Young's excuse for his lack of progress was that he had been ill, but he may have in fact been devoting all of his time to improving his own claim a few miles to the west, using the support of the wealthy Irvines to sustain him.[55] With lawyers roaming the land assuring the landless that they could help them acquire titles by improvement, there were fewer men who were willing to expend their energies improving some land baron's stake. By September, Matt Young was no longer tending the Irvine land but devoting his energies to his own property. Just one year after ending his agreement with the Irvines, Matt Young scrawled the town name "Youngval" on a large rock adjacent to his claim a few miles up the Brokenstraw. No longer would he be a tenant on someone else's land; rather, he envisioned being the proprietor of his own town. Across the region, settlers were claiming lands by right of improvement, including many who had originally been brought there under contract to the Holland Land Company and had accepted provisions and tools from the companies to get established. The Irvines and the other investors in the land companies were outraged by what they saw as the unprincipled behavior of these squatters; for their part, the squatters, emboldened by the language of liberty and equality that filled the air in the post-Revolution years, looked upon the method by which the wealthy had initially acquired claims to these lands as the greater fraud and felt justified in their decision to abrogate their original agreements.

With Matt Young gone, Callender Irvine's land faced additional threats. A neighbor wrote him in September to warn him that "your farm on Brokenstraw is gitting much out of repair. The Indians are burning your Rails, and your Peach trees & other Trees will be exposed to Cattle and everything else." The neighbor offered to take the remaining rails for use on his own property "and I would give you as many as you needed when you might want them. Also, if you would let me have some of your Peach trees I would give you the same number when you commence farming on the Brokenstraw."[56]

Callender remained in government employ for several more years, first as a militia captain at Franklin, and then as a provisions officer at Fort Presque Isle on Lake Erie. Both jobs kept him away from his Brokenstraw lands, but he kept an eye on them thanks to the help of a few friendly neighbors. Nevertheless, challenges to his ownership would continue. As late as 1803 John Andrews's brother Robert wrote to warn him that two men, John Long and Andrew Carr, had been "poking at your premises on Brokenstraw full of expectations that shortly you and the rest of the world will be convinced that the patents will be null and void, they to become the proper owners." This time Callender was unable to drive them away simply by dropping his father's name, and Long and Carr dug in and set about making their own improvements.[57]

Precisely how John Chapman navigated this contentious terrain, and what he aspired to when he first came to western Pennsylvania, can only be pieced together through a thin oral tradition and a handful of documentary traces. The oral tradition offers two different portraits of John during his relatively brief stay in the region. One suggests that he was a victim of his own generosity in a time and place that rewarded the aggressive and self-serving: "He was one of those characters, very often found in a new country, always ready to lend a helping hand to his neighbors. He helped others more than himself . . . he took up land several times, but would soon find himself without any, by reason of some other person 'jumping' his claim."[58] Another version implies that his failure was the result of his own improvidence: "[Chapman] took up land in different parts of [French Creek] township. His sojourn owing to his thriftless disposition, was only temporary . . . he appears to have been impatient with the restraints of civilization, so much so, indeed, that as soon as settlements began to increase he disposed of his few improvements and with a few other spirits as restless and disoriented as himself, drifted further westward."[59]

Both stories seem to recognize that there was something broken about northwestern Pennsylvania society at this time. Together, the two narratives embody the prevailing stereotypes of frontier culture. In the former, Chapman was a symbol of the cooperative frontier-builder, "always ready to lend a helping hand." But in this region wracked by legal fights over landownership, that cooperative frontier spirit was Chapman's undoing. The latter explanation reflects popular negative stereotypes about frontiersmen, that they were thriftless, disoriented, and restless, decivilized by the wilderness. According to this narrative, the transitory and unsettled nature of the region was not the result of bad laws like the Pennsylvania Land Act of 1792; rather, it was an unfortunate but universal stage of frontier development. The first wave of frontier settlers were so negatively affected by the wilderness that they became white versions of the Indians they had dispossessed. These frontiersmen laid the groundwork for farm families to arrive and bring with them the dress, habits, technologies, and values of civilization; then they were either transformed into respectable middle-class citizens or they chased the receding wilderness westward.

If local traditions about John Chapman are framed largely in these popular narratives of frontier culture, do the documentary evidence and surviving stories provide a truer picture of the real John Chapman during his Pennsylvania years? Was the twenty-something John Chapman of northwestern Pennsylvania a thriftless frontiersman, restless of the restraints of civilization, or a young man aspiring to the life of his forefathers—landownership, a home, a farm, and a family? Did he arrive in the region on his own initiative, or was he drawn there with the intention of reaching a deal with the Holland Land Company to settle and improve one of their warranted land claims? Few of the Holland Land Company records survive, but the telltale clues left behind in store ledgers, account books, census returns, and the oral tradition provide some tentative answers to these questions.[60]

Chapman did not settle down quickly in one place after arriving in the region. Ledgers from John Daniels's store on the Brokenstraw, as well as records of the Holland Land Company store at Franklin, demonstrate that between 1796 and 1799, at least, he divided his time between the Brokenstraw and French Creek areas, with no discernable pattern to his movements.[61] Oral tradition tells us that he planted seedling nurseries along the banks of both Brokenstraw Creek and French Creek and that he periodically returned over the mountains, or to the more settled regions down the Allegheny near Pittsburgh, to gather more seeds for new nurseries. His movement between the creeks may have been driven by his need to tend these nurseries and to find potential customers for his seedling trees. Islands in creeks and land along the banks of creeks were his preferred locations for planting, and most of the local traditions across Pennsylvania, Ohio, and Indiana about the sites of Chapman's nurseries place them beside creeks. These locations provided a natural barrier from browsing deer and cattle on at least one side, thus limiting the amount of brush fence he would have to construct. They also provided easy access to water needed for the trees to thrive.

Why Chapman established his nurseries from seed, rather than from budded or grafted stock, would later become the source of much speculation. Many who encountered him in later years attributed the choice to Chapman harboring strange superstitions or religious beliefs. But planting nurseries from seed was a perfectly rational strategy for the frontier. He could acquire seeds for free behind cider mills and carry large quantities of them long distances without harm;

grafted or budded stock was expensive and difficult to procure, and it could only be carried in small quantities. Furthermore, it made little sense to plant expensive grafted stock when many of the young trees would be lost. Flood, fire, weed growth, and browsing wildlife were all natural threats; the fact that he established these early nurseries on lands he did not own meant that some would be claimed by the legal settlers of the land a few years after planting. Given those realities, Chapman's strategy of planting large numbers of trees from seed, scattered across the landscape in hidden nurseries, was a smart one. Those seedlings that survived could fetch a price of between one and six pence each from settlers anxious to get a head start on an orchard.

While the surviving oral tradition from Pennsylvania suggests that Chapman was spending enough time planting apple nurseries that he earned a reputation for this, it was not his only economic activity in the region. On at least two occasions Holland Land Company agents paid off part of his debt at the company store, presumably in exchange for his labor, and on a third occasion Chapman was paid for four days labor for driving cattle to the Holland Land Company mill at Oil Creek.[62] And there was the tradition mentioned above that suggested that John may have worked as a logger. More revealing is his appearance at Alexander McDowell's Fort Franklin store in May and again in June 1799, each time bringing three bushels of corn for sale. This suggests that in the previous summer he had established a land claim somewhere nearby and had planted and harvested corn the previous year. It also implies he had use of a pack animal, as he could not have carried three bushels on his own back. Finally, his economic activities with Holland Land Company officials indicate that he was on good terms with them and that the company did not see him as a squatter intent to lay claim to lands they held by warrant, as the company routinely refused to sell goods or do business with persons officials believed were there to steal from them.[63]

The Brokenstraw store ledger also affirms an oral tradition claiming that John's younger half-brother Nathaniel joined him in the region. Nathaniel appears in the documents in June 1798, paying for goods "by work."[64] The appearance of Nathaniel after a winter and spring when John was absent from store ledgers in both Brokenstraw and Franklin suggests that John visited family in Longmeadow that winter and brought young Nathaniel, just sixteen years old at that time, back to Pennsylvania on his return trip.[65] John's continued contact with his family back east also suggest that he did not leave for the West intent on a life of hermitage but probably with much more conventional ambitions. Perhaps he was sent west with the blessings of his family, to scout out suitable places for their eventual migration.[66]

FRANKLIN: THE FRONTIER BACHELOR

During the years John Chapman was trying to build a life for himself in the upper Allegheny watershed, the region's other bachelor apple-tree planter, Cornelius Van Horn, was doing the same. In 1795, the majority of the whites in the region were solitary men—either bachelors or husbands who left families behind until it was safe to bring them. But with the threat of Indian attacks gone and his land title secure, Van Horn decided it was time to make a change: "From the year 1795 to the year 1798 I kept house as a bachelor nearly three years, when I concluded to give up house-keeping or agree with a woman to be my housekeeper. Sarah Dunn was the first I applied to who agreed with me and we were married the 27th day of September 1798."[67]

With this rather unromantic agreement transacted, Van Horn completed the transition from frontiersman to yeoman farmer. He and Sarah began to live the life the Quaker missionaries had tried to convince Cornplanter's people to embrace. Cornelius devoted his days to cultivating fields and orchards and tending livestock, and Sarah attended to domestic duties. Children would follow. Van Horn spent the rest of his life along French Creek, in old age, taking on the role of founding pioneer and coming to the attention of those who sought to record the history of the region's earliest days.

John Chapman, by contrast, eventually departed the region, leaving behind only traces of his presence in yellowing account ledgers, some apple trees, and a few yarns in the collective oral tradition of northwestern Pennsylvania. The question of why he left remains. Was he, as one local tradition recounted, simply another of these restless, uncivilized frontier characters with a "thriftless disposition" who retreated west as Pennsylvania's forests became farmlands? Or was he, as a more charitable recorder concluded, a victim of his own selfless and generous spirit, taken advantage of by more aggressive pioneers who jumped his claim? The evidence suggests neither characterization is accurate.

While John Chapman appeared to move regularly between his nurseries along the Brokenstraw and those he had set out along French Creek during his first few years in the region, after 1799, Franklin and the lands below French Creek appeared to be his base of operation. His prospects for selling apple seedlings to the very few farmers in the northern reaches of Holland Land Company lands were quite limited, and securing a land claim in that region, whether he signed a land contract with the Holland Land Company or not, meant being prepared to fight to keep it. Warren and the Brokenstraw country were the back door to the Holland Land Company purchases. Franklin was the front door. It could be reached from the Wyoming Valley by a southerly route over the Alleghenies, and it was also easily accessible by water or trail from Pittsburgh.

Franklin had more promise than Warren, but it was by no means experiencing a settlement boom. In 1795 when surveyors plotted a town center and began laying out streets, Franklin was nothing more than a small military outpost, "but five or six little dirty log huts surrounded by a great wilderness of seventy or eighty miles with Indians hooping and halloing and begging for whiskey," according to one member of the survey party. A store ledger from Franklin in that year records a customer base that was almost entirely Indian.[68] Nevertheless, with the Holland Land Company expecting to settle the land in the next year, the surveyor speculated on the changes coming. "This is indeed very unlike Philadelphia but perhaps in process of time the howling desert may be turned into pleasant fields, and shining bricks decorate this spot of ground which now appears so unlike a city of commerce."[69] Its location at the confluence of French Creek and the Allegheny, and its role as a launching point for settlement of Holland Land Company tracts ensured Franklin's modest growth. By 1800, the village contained thirty-eight households, and the white population for Venango County was 1,130.

The potential for agricultural settlement in Venango County was mixed. Cornelius Van Horn and the other settlers who used Fort Franklin as a base for their settlements poled up French Creek past the lands nearest Franklin and instead traveled twenty miles upriver to flatter, more fertile lands in what eventually became Crawford County. Near Franklin, French Creek descended through a steep gorge, with only intermittent bottom lands along its sides. On the north side of the creek, the lands showed some promise as farmlands, and the Holland Land Company acquired most of these. The south bank of French Creek appeared less desirable as farmland. For eight miles up the river from Franklin, steep forest-covered hills rise up from the along French Creek's south bank. Today, several pleasant small farms are carved out between the woodlands on the flatter pieces of ground high above French Creek, but they are separated by large tracts of forest. William Irvine had described the region below French Creek as having "no land worth mentioning fit for cultivation . . . it is a continued chain of high barren mountains except small breaches for Creeks and Rivulets."[70] As a result, the land barons, who had all carefully read Irvine's report, left these lands alone.

It was in these unwanted lands south of French Creek that John Chapman first settled. Chapman allegedly took up several different plots of land in this region but kept finding himself dispossessed by others jumping his claim, and the provisions of the 1792 Land Act were certainly easily exploited by the aggressive and unscrupulous. The census of 1800 lists him as living in this area as a bachelor. If John Chapman limited his efforts to claim land to these unwanted lands, and all of the surviving evidence suggests he did, it also explains how he managed to remain on good terms with Holland Land Company officials without signing a land contract with them. The fact that some of his many primitive nurseries were planted on Holland Land Company land would not concern them so long as he made no attempt to build a cabin or lay out other improvements, and company officials probably saw his efforts to bring orchards to settlers as aiding their efforts to settle the lands. Despite Irvine's pessimistic reports of the lands in the Sandy Creek region, they were populated more quickly than the lands to the north, which were mostly controlled by the Holland Land Company. It appeared that John Chapman, and a majority of the early settlers preferred to try acquiring land titles by right of improvement rather than entering into a contract with the land companies, even if it meant settling for less desirable lands.[71]

John Chapman's inability to secure a claim was the most obvious barrier he faced to following in Van Horn's path to domestic bliss. For five previous generations of Chapmans, acquiring land or at least a competency—skills and the tools needed to employ them profitably—were the prerequisite conditions for marriage. As long as Chapman was unable to secure title to a piece of land he was simply not marriageable material. Perhaps the eccentricities which he became known for in later life were already present, and also made him a less desirable candidate for matrimony. The surviving store ledgers provide only a few insights into Chapman's personal characteristics. The "two small histories" John purchased at the Brokenstraw store in the spring of 1797, at a time when he may have just emerged from a winter holed up in the hollow of a Sycamore tree are the only books, save a few Bibles, sold to any customer at this primitive outpost in the last five years of the eighteenth century. If literacy and a thirst for knowledge were things that stood him out from other frontier bachelors, they hardly seem like distinctions which would be a liability in the competition for a mate. Most of the entries, including one from July 1797, suggest John Chapman was not so different from his frontier neighbors and undermine the static myth of "Johnny Appleseed" as a barefoot, clean-living vegetarian who never carried a gun. Among the items he purchased that day: brandy, whiskey, tobacco, pork, chocolate, sugar, gunpowder, and several pairs of moccasins.[72]

Despite the continuing harsh conditions in northwestern Pennsylvania, the rapidity with which bachelorhood disappeared from the region was striking.

By 1800, only one out of six households in Venango was occupied by solitary males. John's continued bachelor status at age twenty-six was not unusual, but in the next decade it became increasingly rare. By 1810, by which time John Chapman had departed the region, only about one in twenty households in his old neighborhood was occupied by single men. After 1800, despite its continued troubles with unresolved land claims, northwestern Pennsylvania was rapidly becoming domestic. Single men living in rough quarters on the edges of settlements confronted Cornelius Van Horn's choice—either find a wife and settle down or move west with the frontier.

One of the results of the rapid domestication of Franklin and Venango County was that bachelors, especially those who had arrived when the region was still a frontier, became objects of fascination, particularly for children who yearned to hear the tales of wilderness adventure. It appears that John Chapman was willing to play the role of rugged frontiersman and share stories with the increasing numbers of children he encountered during his last years in Pennsylvania. Many of the first-person recollections of Chapman that survive come from people who remembered him from childhood, including the only first-person account from Pennsylvania. R. I. Curtis, who spent his childhood along French Creek, recalled encountering Chapman when he was a young boy in the first years of the nineteenth century. For the young Curtis, the rugged Chapman was a noteworthy contrast to the men like his father who had settled into a more domestic existence. Curtis did not publish his memories of Chapman until old age, by which time the myth of Johnny Appleseed had begun to circulate in regional newspapers and magazines, and it is clear from Curtis's account that he had read many of the Ohio stories about Johnny Appleseed. Some of Curtis's recollections affirm elements of the Appleseed lore, including his kindness toward children and his Christlike meekness, for example. But a different Chapman, one much more in line with the dominant frontiersman myths surrounding figures like Davy Crockett and Mike Fink, also appears in the account. Curtis recalled Chapman as a man of extraordinary strength, who "could chop as much wood or girdle as many trees in one day as most men could in two." And the stories Chapman told Curtis and the other children contained more than a small dose of frontier braggadocio. In one tale, Chapman claimed to have escaped some menacing warriors by submerging himself underwater in a marsh, lying flat on the muddy bottom while breathing through a reed. He spent so much time underwater waiting for the threat to pass that he fell asleep, awaking, still submerged, several hours later. In another story John Chapman was navigating one of the region's waterways by canoe in later winter while the river was still filled with treacherous, boat-crushing ice floes. The cool-headed Chapman simply dragged his canoe onto one of the larger cakes of ice, climbed back in the boat, and promptly drifted off to sleep, awakening one hundred miles south of where he intended to disembark.[73] These descriptions of extraordinary calm, even indifference, when facing life-threatening danger are a common element of frontier folktales and are echoed in Van Horn's apple seedling nursery tale.

But one Curtis story, which depicts Chapman as a man of extraordinary meekness, seems to reinforce the argument that he was unable to secure a land claim because of his unwillingness to scrap for what was his. "He never resented an injury. I once saw him most outrageously abused by a man much smaller than himself, for some offense that he had unwittingly committed. After reviling him in the hardest language, he kicked him, all which Johnny bore with great meekness, and totally unruffled, and if he had been struck on one cheek, he would have turned the other."[74]

John Chapman's departure from northwestern Pennsylvania appears to have come in stages. He was spotted in Ohio as early as 1801, but the last document affirming his presence in Franklin, Pennsylvania, was in December 1804. An oral tradition has him returning from a trip to Pennsylvania through eastern Ohio in 1806.[75] In all probability, he spent several years moving back and forth between Ohio and Pennsylvania, returning to check on his maturing orchards in the east and gathering seeds from Pittsburgh-area cider mills and carrying them back to Ohio where he began planting more nurseries. The prevailing oral traditions in Pennsylvania interpret his move west as evidence of failure, indicating that he was the victim of aggressive claim jumpers or his own thriftlessness. The contentious nature of the scramble for land in the region, combined with Curtis's description of John Chapman's meekness, lends some credence to the theory that he was the victim of claim jumpers. But the notion that it was thriftlessness and lack of ambition that brought about his departure seems less credible. Topography, politics, and economics were limiting northwestern Pennsylvania's capacity for population growth. The region's arable farmland was restricted to its flood-prone river bottoms. The thin, stony soils of its many steep ridges held out fewer prospects for farm families. Many of the region's early settlers abandoned farming for lumbering, clear-cut plots to which they had no legitimate title, floated the lumber down the Allegheny to markets in Pittsburgh, then "took French leave," moving on to new plots.[76] Furthermore, one of the effects of this revolt against the land barons and the ensuing legal contest was to suppress settlement and development in the region for years, as potential settlers quickly decide to move further west, into lands in Ohio, where acquiring a title seemed an easier task. The effects of the unresolved legal conflict depressed development in the region for decades. In 1813 the town of Warren still contained just five houses. As late as 1816, a traveler passing near Franklin noted, "An almost total want of energy prevails. Many small farms, which had been cleared some years ago—including fruit trees of handsome growth— are completely deserted . . . even the improvements of former years, now occupied, are retrogressive. The chief part of the remaining inhabitants reminded us of exiles; and if they escaped the ravages of famine, they must be nearly estranged to the common comforts of civilized life."[77] As a result, the market for Chapman's seedling trees in the upper Allegheny watershed was limited, but a great demand for apple trees existed in the upper reaches of Ohio's Muskingum Valley.

John Chapman spent roughly a decade of his life in Pennsylvania, In his young adult years—essentially his twenties—he left only a scant record of his travels and activities. But his time in Pennsylvania was critical to his formation. It was in Pennsylvania that he developed the skills he needed to survive in the wilderness and the capacity and willingness to live in the most primitive conditions. It was here that he planted his first apple seedlings and offered them up for sale to pioneers, an activity that would eventually come to define him. Yet in Pennsylvania he was still John Chapman, not yet "Johnny Appleseed." He was a young man, much like the other young men who appeared there in the first years, who aspired to the independent manhood that land ownership and a self-provisioning family farm might secure. In the traces left behind in land company records and the thin surviving strands of the region's oral tradition emerge glimpses of the famous Johnny Appleseed he would become after steering his canoe down the Allegheny past Pittsburgh for the last time and into the Ohio country.

CHAPTER THREE

Suckers

One sign of an untended apple tree is the appearance of suckers. Suckers, also called root sprouts, water sprouts, and basal sprouts, are the thin branches that rise from the soil at or near the base of a tree. Fruit trees are prone to suckering, and a diligent orchardist regularly checks her trees for the emergence of suckers, clearing them from the roots as they appear. But the desire of the sucker to rise to the sun is stubborn, and many orchardists wage a continuing war against them. Each orchardist has her theories and strategies. Remove them by tearing, not cutting, and they are less likely to come back, one will tell you. Wait until July to remove them, when the torpor of summer encourages the sucker to surrender, defeated, another advises. Most modern orchardists see suckers as a disorderly nuisance, an unproductive competitor for soil, water, and sun with the tree that spawned them. Let suckers be, they will warn you, and they will threaten the vigor and productivity of your orchard.

With care, however, a sucker can be removed from its parent tree, replanted, and grow into a tree in its own right. On modern trees, any sucker that emerges from below the graft will not produce fruit like that of the main tree but, instead, like that of the rootstock. Rooting such a sucker is a lottery. The fruit that emerges from the tree is most likely to be gnarled, crabbed, and bitter, but every so often an apple of remarkable qualities emerges from a lowly sucker. Some of the most popular varieties that survive today were born of a disdained, unwanted sucker.[1] Yet most modern horticulturists will advise the orchardist to remove and discard these pests.

On a natural, ungrafted tree, a transplanted sucker will produce fruit of the same characteristics of its parent. In the early days of American orcharding, if a farmer discovered a natural tree in his orchard or field with superior fruit, he might propagate it by gently pulling the sucker away from its parent and rooting it in open soil. When tens of thousands of Eastern farm families crossed the

Appalachians into the West in the early nineteenth century, some of them carried with them a few suckers from their favorite tree from home.[2] Others identified suckers with untidiness, disorder, and idleness.

Philadelphia physician Benjamin Rush associated unproductive suckers with unproductive people and believed that both needed to be removed for trees and communities to thrive. In an essay on the progress of population and agriculture in Pennsylvania published in 1798, Rush warned that overpopulation in eastern Pennsylvania threatened the health of agricultural communities. As farms became subdivided into smaller parcels when they passed from generation to generation, the possibility of prosperity for anyone in the community was threatened. Farms of seventy-five to three hundred acres were ideal, Rush believed, and as population increase in the rural East reduced average farm size, "a languor in the population" was the result. The solution, Rush concluded, was removing the excess population to the West—specifically "the idle and extravagant, who eat without working." Once this was done, the industrious who remained would thrive once again, "just as the cutting off the suckers of an apple-tree increases the size of the tree, and the quantity of fruit."[3] Benjamin Rush was not the only person to compare the rural transient poor to suckers; one of the early nicknames of Illinois was the Sucker State, derived, according to one of its early governors, from an epithet bestowed on the waves of poor Southerners who migrated there "because they were asserted to be a burthen upon the people of wealth; and when they removed to Illinois they were supposed to have stripped themselves off from the parent stem and gone away to perish like the 'sucker' of the tobacco plant."[4]

MARIETTA

The Chapmans were suckers. In the years after Shays's Rebellion, they ground out a living as tenant farmers, struggling to feed a growing family with meager resources. As poor tenants in Longmeadow, they left only the lightest traces on the historical record. Recovering the story of the Chapman family requires carefully overlaying family traditions with scattered census records and genealogies and a little bit of speculation. Lucy Cooley gave birth to ten children between 1781 and 1803, five boys and five girls, although one of the girls probably died in childhood.[5]

The prospects for landless farmers in the Connecticut Valley did not improve in the years after Shays's Rebellion, and farm sizes continued to shrink as they had been for generations. Most New England farmers in these years had to supplement their farm income with other work, and Nathaniel, trained as a carpenter, was probably no exception. The formation of the Ohio Company and Associates by a group of Revolutionary War officers was one response to the problem of overpopulation. The Ohio Company secured a large grant of land from Congress and sold shares to those who were interested in planting a new Yankee colony on the north side of the Ohio River. In 1788, a group of Ohio Company shareholders gathered at Longmeadow and set off en masse for the new colony on the Ohio at the mouth of the Muskingum River. The carefully platted town of Marietta was to be the seed of an ambitious effort to transplant Yankee culture to new lands in the West. As the Ohio Company pioneers gathered for their departure, we can imagine that many Longmeadow families, the Chapmans among them, looked on with curiosity and perhaps some envy. But Nathaniel Chapman, struggling to feed a growing family, could not afford to join them.[6]

Events in Ohio swiftly threatened to undo the ambitious plans of the Ohio Company. Ohio's Indian peoples were not prepared to surrender their land to the newcomers. When a small group of Yankees began erecting cabins and overturning the earth at Big Bottom, twenty miles north of Marietta along the Muskingum, they were set upon by Delaware and Wendat warriors, who killed nine men, one woman, and two children. Shareholders in Marietta and back east were worried. Nothing stood between Big Bottom and the town of Marietta, and after the killings, few white families with resources were eager to settle its hinterland. The associates' solution was to persuade Congress to grant them a "donation tract" north of the town and offer up hundred-acre parcels to "warlike Christian men" who were willing to hold their claim with both rifle and plough. These men were required to acquire arms and a ready supply of ammunition and, within three years, build a cabin, clear fifteen acres of land for pasture, and set out an orchard with at least twenty peach trees and fifty apple or pear trees. When news of the donation tracts reached Longmeadow, Lucy Chapman's Cooley relatives responded. Asahel, Reuben, Jabez Jr., and Simeon, sons and grandsons of her uncle and aunt, Jabez and Abigail Cooley, set out for Ohio to make claims.[7]

By the time the Cooley boys arrived in Marietta in 1792 or 1793, the company had a large nursery of seedling apple and peach trees ready for transplanting to new orchards on the frontier. In 1794 Israel Putnam returned from the Connecticut Valley with several saddlebags filled with apple grafts, carefully coated in beeswax to keep them alive. Within a few years, cuttings from his Westfield Seek-No-Furthers, Rhode Island Greenings, Roxbury Russetts, and Late and

Ohio in 1800. Claudia Walters, University of Michigan–Dearborn; adapted from a map drawn under the supervision of Frances P. Weisenburger.

Early Chandlers were available for grafting onto the Cooleys' maturing seedling apple orchards if they so desired.[8]

Late that same year the Indian confederacy collapsed in defeat at the Battle of Fallen Timbers, and the threat to Marietta passed. In the spring of 1797, the Ohio Company relaxed the five-year residence requirement on donation tracts, and all four Cooley boys were granted outright titles to their lands. Prospects for Ohio lands were looking up, and migration to the region accelerated. The landed Cooleys were soon buying more land and selling it to the newcomers.[9]

Nathaniel and Lucy Chapman had been in no position to take advantage of the donation tract offer in 1792. Even if the land were free, moving a family west required a significant outlay of money and strong hands and backs to carve a homestead out of wilderness. Nathaniel Chapman's eldest son, John, turned eighteen that year, but his five younger children, between the ages of two and eleven, would be of little help. The extended Cooley clan had several adult men up to the task, but the Chapmans stayed put.

As we have seen, John eventually left for northwestern Pennsylvania, where he was briefly joined by his half-brother Nathaniel in 1798. One local tradition asserts that Nathaniel Jr. first appeared in the Marietta area as early as 1798. It is not hard to imagine that Nathaniel and his brother John both traveled there that year to visit their Cooley cousins and check out the region's prospects. Whether Nathaniel Jr. appeared in Marietta that early or not, by 1800 the census revealed that two of the Chapman sons were missing from the Longmeadow home—almost certainly Nathaniel Jr., nineteen, and Parley, fifteen. It appears that Nathaniel Jr. returned the favor his brother John had done for him and took his younger brother off to the West. Son Abner, seventeen at the time, appears to have remained behind to help his father with the Longmeadow farm. Marietta traditions mention that it was Nathaniel and Parley who guided the family to a new homestead in Ohio, so it is reasonable to speculate that Nathaniel Sr. sent the two young men to prepare the way for an eventual family migration. In the meantime, John continued to be a presence in northwestern Pennsylvania but had also pressed into Ohio and begun planting apple seed nurseries there.[10]

In the winter of 1804–1805 the Chapman family finally made the move west. By this time, John was thirty years old, and Nathaniel Jr., Abner, and Parley were between the ages of nineteen and twenty-three. Nathaniel Sr.'s daughter Lucy, seventeen, was old enough to bear part of the burden. But the household also included five younger children ranging in age from one to eleven. The Chapman family's decision to leave Longmeadow may have been planned for several years, but it also was the result of desperation. Fifty-eight-year-old Nathaniel Sr. appeared to be in poor health. Two mysterious bills of exchange that John Chapman signed in Franklin in December 1804 were probably connected to his family's impending migration. The first promised to repay, with interest, one hundred dollars to the children of his sister Elizabeth, who was still residing in Massachusetts. The second promised to pay one hundred dollars "in land or apple trees" to a Nathaniel Chapman, probably his brother rather than his father. John had gone to some effort to make these notes fit the conventions of legally binding bills of exchange rather than less formal promissory notes, which neighbors and family might offer one another. He employed the common terminology of the bill of exchange, indicating these debts were "for value received." He went to the trouble of having each bill signed and dated by two witnesses. But John, inexperienced in the rules of formal financial transactions, didn't quite get it right. Both documents left out a critical piece of information—when the payment came due—making them of questionable legal enforceability.

But why exactly did John issue these bills to the family of his sister and to his brother in December 1804, at precisely the time his struggling Longmeadow family was packing up to move west? It is unlikely he was actually borrowing money from them. A more likely explanation is that this was John's attempt to meet a filial obligation at a time of family crisis. Moving a large family west cost a substantial amount of money in supplies. John was not in a position to pay part of that cost, but his older sister's family, established back in Charlemont, and his younger half-brother Nathaniel, by this time set up in the Marietta area for perhaps as long as five years, had covered the costs. In constructing these almost-legal bills of exchange, John wanted his sister and brother to have more than his word that he recognized their sacrifices and would at a later date compensate them for their shares. The language "for value received" gave them more legal weight, but does not in fact mean he actually received anything in exchange for these promises to pay. There were two other curious elements of these bills. In the first he promised to pay with interest the amount due to his sister's two children, Nathaniel and Elizabeth Rudd. He must have not received a communication from his sister for several months at the time he signed these, as both children died of illness a few months before John Chapman issued this note. His sister Elizabeth named her next born John, after her brother. In the second note, John promised to make the payment "in land or apple trees." This indicates that by 1804 he had come to see apple tree planting as a primary means of earning a living and also that he intended to acquire land. This information also reinforces the theory that he sent the note to his brother, not his impoverished and ill father, who would have little use for either forms of payment down the road.[11]

The Chapman family arrived in Ohio early in 1805 and settled on lands along Duck Creek in the donation lands or just north of them. Nathaniel Sr. never acquired title to any land, so it is probable that the family squatted on one of the many unclaimed parcels there. Some of these tracts had been abandoned or never taken up, and Congress did not ask the Ohio Company for an accounting of them until 1818, giving squatters an opening. The lower Duck Creek Valley is a land of steep slopes and narrow, flood-prone bottom lands, making it some of the least desirable land open for settlement. The Chapmans may have selected this land precisely because they had no money and they expected to be left alone there.[12] Another possibility is that the Chapmans settled on one of the Cooley tracts in the area. Oral tradition and the 1810 census disagree about the precise location of the Chapman family homestead, but both place them somewhere along a ten-mile stretch of the steep, muddy hollows of Duck Creek.[13] Whether the Chapmans were squatters or were charity tenants on a Cooley tract in these early years, it appears that the family was barely getting by, eking out a subsistence on some of the most isolated and unpromising lands in the region.

The property was seven miles from the nearest neighbor, according to one early account. Just two years after they arrived, Nathaniel Sr. passed away, and the family was so poor that they could not afford a proper burial for their father. Nathaniel Jr. and Parley walked seven miles to the nearest sawmill at Salem, acquired a few cut boards, and borrowed the tools needed to make a rude coffin, which they carried back to the homestead. They buried their father in an unidentified grave somewhere on the property.[14]

Nathaniel Sr. never achieved the independence that his forefather Edward secured for his children, and he died a pauper. His widow, Lucy, did not fare much better. The 1810 census reveals her living along Duck Creek with just her two youngest children, Davis and Sally. Percis, Mary, and her son, Jonathan Cooley, had probably been parceled out to live with other family members. But Ohio would offer Lucy and Nathaniel's children a better life. The year after their father died, both Nathaniel and Parley married, and both eventually became landowners and had sizeable families. Decades later, Parley achieved enough respectability in the area to be elected to serve as the grand marshal of his local Odd Fellows lodge.[15] His grave sits on a hill rising from the Duck Creek Valley on property once owned by his wife's family. Nathaniel Jr. was residing with his young wife about fourteen miles west, on lands near the Muskingum River. He eventually built a mill along Olive Creek at Moscow Mills and acquired several parcels of farmland in Washington and Morgan County.[16] Abner Chapman, living alone in the same township as his mother in 1810, may have moved further downriver to Gallipolis. Jonathan Cooley Chapman—according to family tradition a deaf-mute—lived alone near Broken Tree a few miles west of Duck Creek. Sister Percis married a local man named William Broom, and the couple eventually migrated to central Ohio. Baby sister and brother Sally and Davis also married in the region and are buried nearby.[17]

John Chapman had left his family in the late 1790s. Their arrival west in 1805 gave him an opportunity to reunite with them. But while he remained close to his siblings and visited them regularly, he did not settle among them in the Marietta region. In the popular imagination, Johnny Appleseed is portrayed as a lone wilderness wanderer, a man with few attachments. This is largely untrue. He visited his Marietta family regularly and maintained close ties to some of them for the rest of his life. But by the time his family made the move to Ohio, he had apparently settled on a life plan that required him to go elsewhere.[18]

ALONG THE TRACE

John Chapman's interest in Ohio went beyond concerns for the welfare of his family. While the settlements of northwestern Pennsylvania stagnated as a result of the contentious legal issues surrounding land titles, migration to Ohio continued to grow. For a man determined to make a living planting and selling apple trees, Ohio offered great promise. But not every part of Ohio was fertile territory for his seedling nursery business. In particular, the region around Marietta was not promising. Ohio Company lands near Marietta had plenty of mature orchards of both seedling and grafted stock by 1805. There was little need for his seedling trees. Local traditions suggest he planted a few nurseries here, or even helped one resident set out an orchard, but there is no evidence that John engaged in sustained economic activity in the region.[19]

To the northeast of Marietta, on the lands known as the Seven Ranges, bordering the Ohio River, we hear of his first trip into Ohio with apple seeds. According to this tradition, John crossed the Ohio River from the Virginia side in the spring of 1801, leading a packhorse loaded down with small leather bags filled with apple seeds, seeds he had gathered from Pennsylvania cider mills the previous fall. The settlements on the Ohio side of the river were still quite primitive in 1801, but lands were filling rapidly. John Williams, then just eleven years old when his family migrated from North Carolina, described the scene in 1800 this way: "Emigrants poured in from different posts, cabins were put up in every direction, and women, children, and goods tumbled into them. The tide of emigration flowed like water through the breach in a milldam. Everything was bustle and confusion, and all at work that could work."[20]

There was much work for John Williams, his widowed mother, older sister and brother to do. Building a primitive shelter was the most essential first task. Before corn could be planted, the thick-trunked beech trees that covered the claim had to be girdled. Smaller brush had to be cleared and the soil scratched up. "We were weak-handed and weak-pocketed—in fact, laborers were not to be had," Williams recalled, and the hands of every member of the family were needed just to get by. In August the family finally found time to plant turnips, and in the fall they spent time in the woods gathering walnuts and hickory nuts. "These, with the turnips, which we scraped, supplied the place of fruit."[21]

The family planted some peach stones as well, as peach trees grew fast and bore fruit in just a few years. But in those first years, sweetness could only be found in a turnip or a walnut, or for the moment each year when the delicate pawpaw passed through its brief transition from ripeness to rotting. When a family's diet consisted almost entirely of milk and corn mush, how must these items have tasted? By the time John Williams was sixteen, he no longer had to do with such limited sweet tastes; stone-grown peaches were so abundant that "millions of peaches rotted on the ground." Eventually the Williams family found the time to plant apple trees, but in the meantime, the tart-sweet tang of an apple occasionally could be had, for a boy willing to work for it. "We sometimes went to Martin's Ferry, on the Ohio, to pick peaches for the owner, who had them distilled. We got a bushel of apples for each day's work in picking peaches. These were kept for particular eating," John Williams recalled, "as if they had contained seeds of gold."[22]

Eastern Ohio appeared to be a perfect place for John Chapman to set out seedling nurseries. And when he arrived in the area in the spring of 1801 with his seeds, he attracted some attention. One local pioneer who was pleased to learn of John Chapman's mission offered him a place to stay overnight and spent much of the evening trying to persuade John that his seedling trees would be greatly valued in those parts.[23] But John did not tarry there. Perhaps he recognized that this moment of apple deprivation would be gone before any seedlings he planted in 1801 were ready for transplanting to orchards. Just across the river in Wellsburg, Virginia, Mennonite Jacob Nessly, who had arrived from Lancaster, Pennsylvania, with his family some fifteen years earlier, already had fifty acres of maturing trees. Most of these were seed-grown trees, whose fruit was distilled into apple and peach brandy, which Nessly sold to downriver markets. But by the time Chapman arrived, Nessly was already offering young seedlings for sale and had begun top-grafting some of his seedling trees with winter apples, which he would soon have for sale to markets as far away as New Orleans.[24] And just twenty miles downriver, Ebenezer Zane had established some orchards on Wheeling Island, a stone's throw from Martin's Ferry, where John Williams was soon picking peaches for another local distiller for a bushel of apples a day. As apples became more abundant in the region, the cost of a day's labor, measured in apples, no doubt escalated. If apples were still scarce on the western side of the Ohio River in 1801, as a few residents recalled, they did not remain scarce long.²⁵ Soon apple harvests eclipsed those of the short-lived peach. John Williams reflected in his memoir that "when we got an abundance of apples they seemed to lose their flavor and relish."²⁶

John Chapman declined his host's request to stay and plant his nurseries. He thanked him for his hospitality and begged leave after asking for directions to Zane's Trace.²⁷ He may have planted a nursery near the headwaters of Stillwater Creek, not far from the Trace, and another further north, in present-day Carroll County, as local traditions claim. But most of the surviving traditions concerning John Chapman in the Seven Ranges are stories about his passing through. By 1806, when he was seen floating down the Ohio River in two canoes lashed together, filled with apple seeds from Pennsylvania cider mills, eastern Ohio had no need for his seeds. He was headed down to Marietta, no doubt to visit family, before dragging and poling his cargo up the Muskingum River and its tributaries into the Ohio interior.²⁸

After leaving eastern Ohio, John Chapman probably headed west on Zane's Trace. By 1801 there were about fifty thousand settlers living in the ill-defined territory between the Ohio River and Detroit. Over the next decade, twenty thousand new settlers arrived every year, leading to a five-fold increase across the decade.²⁹ The paths and traces John Chapman trod as he traveled between Pennsylvania cider mills and his isolated nurseries grew smoother and wider each year as new migrants moved along them. What had begun six years earlier as a narrow path, once described as "a tight fit for a fat horse," carried mail service by 1798. By 1801, some of the more tenacious migrant families pushed and pulled wagons along the rutted road. In 1805 conditions along the road were still quite dreadful but just adequate for a stagecoach service.³⁰ Driving a wagon through the muddy ruts of Zane's Trace was no easy task, but that did not deter waves of land-hungry settlers.

Every place John Chapman visited during his first decade in Ohio was different from what it had been one year before, and from what it would be the year following. Newcomers arrived, and more "women, children and goods . . . tumbled" into primitive cabins. Others moved on, but almost everywhere the direction of change was more people, not fewer. Girdled old-growth trees slowly withered and rotted, and more corn rose in the spaces between them. Hogs multiplied and grew wild and fat on the mast of the forest. One-room log cabins were expanded, then clapboarded into respectability. Forest gave way to field; oak, hickory, and beech giants disappeared to be replaced by young apple, peach, and pear trees, which soon needed to be fenced to protect them from hungry, half-wild hogs. The moment between fruit scarcity and fruit abundance was fleeting. Timing was everything for John Chapman's seedling nursery plan. Two- to three-year-old seedlings needed to be nearby when the frontier family had freed up just enough labor to clear a field and plant them, after their stomachs were filled with cornmeal mush, scraped turnips, and the wild nuts of the forest and their palates began to rebel and demand real sweetness.

Zane's Trace was just one route taken by the streams of settlers pouring into the region. It carried settlers from Pennsylvania and New Jersey, mostly of German or Scots-Irish origin. Some upland Southerners would also make use of it, but it was more common for the latter group to travel by flatboat or pirogue down the Ohio River from Wheeling, Virginia, then up the Muskingum, Hocking, Scioto, Little Miami, or Great Miami rivers into the Ohio interior. As John continued down the Trace through the hills of eastern Ohio, he passed growing settlements of Scots-Irish Pennsylvanians. At Zanesville, where the Trace crossed the Muskingum, Chapman found a young community occupied by a mix of Pennsylvanians, Virginians, and Marietta Yankees. West of Zanesville the hills began to flatten, and the Pennsylvania Dutch travelers who had no desire to farm the steep hills of eastern Ohio got busy carving out farms, building their trademark hand-hewn colossal barns and laying out compact villages of brick I-houses and distinctive diamond-cut town squares. By the time the Trace crossed the Scioto River at Chillicothe in the Virginia Military District, upland Southerners were carving irregularly shaped farms out of the landscape, using the colonial metes-and-bounds method, with boundaries following natural ridges and stream beds, in defiance of the rationalized grid Congress had sought to impose across the entire region.

In these Virginian-dominated lands of the lower Scioto River valleys, an emerging Virginia elite, represented by men like Thomas Worthington, Nathaniel Massie, and Edward Tiffin had acquired large landholdings and were selling them off in smaller parcels to poorer Southern migrants. These Virginia gentry, who dominated the early politics of the state, brought with them the apple culture practices of their idols, Thomas Jefferson and George Washington. Shenandoah Valley's Thomas Worthington had inherited land and slaves from his family but believed he could do better for himself by uprooting to Ohio. After making two exploratory trips to the region in 1796 and 1797, the second with his brother-in-law and fellow Virginian Edward Tiffin, the two men secured claims to a substantial amount of land in the region, and each built a primitive cabin in the village of Chillicothe to secure the lands from squatters. The following March, Worthington and Tiffin, both destined to be governors of the future state, gathered their families and their many possessions in a long train of wagons and set out for Ohio.

In contrast to the Williams family, who arrived in the Seven Ranges lacking even some of the most basic tools needed to build a farm, Worthington's wagon train contained not just necessities and every possible tool and implement required to transform his property into a civilized estate but also plenty of luxuries and symbols of his standing. Two large and fragile pier-glass mirrors made the journey, as well as a load of grafted fruit trees. Worthington freed his slaves before entering Ohio, where slavery was illegal, but he kept them on as laborers. He was neither "weak pocketed" nor "weak-handed" as the Williams family had been, so it did not take him long to transform his primitive cabin into a comfortable log dwelling, "with about 300 apple trees, and a great number of peach [and] plum trees" in the yard. Yet even this town residence was not adequate for a man of his ambition, and he soon set about building a large estate on a nearby hilltop, surrounded by gardens and orchards and suitably named "Adena," to emphasize its grandeur. While Thomas Worthington had done much of the grafting and pruning of his first orchard himself, at Adena he emulated Jefferson again and hired German Redemptioners to tend to his gardens and orchards. Soon Worthington was sharing his fine fruit, and even grafts from his trees, with his many distinguished guests, who no doubt marveled at the contrast between his stately home with its gardens and orchards and the primitive cabins and only partially cleared land still inhabited by his poorer neighbors.[31]

Other Virginia elite who immigrated to the region did the same. Edward Tiffin also soon had orchards of fine grafted fruit. Nathaniel Massie had at his disposal thirty men and thirty ploughs to quickly put three hundred acres of land under cultivation.[32] When fellow Virginian William Henry Harrison was appointed governor of the Indiana Territory in 1800, he quickly set about building an Adena-like estate at Vincennes, which he dubbed Grouseland. Although Harrison struggled to pay the bills to complete Grouseland from his family inheritance and modest government salary, the home and grounds were eventually completed and surrounded by gardens and an orchard of fine fruit trees, perhaps from grafts secured from his friend Thomas Worthington.[33] Well-tended orchards of fancy and rare fruit trees had been a status symbol for Thomas Jefferson, a sign that he was a person of substance and refinement. They also revealed that he possessed substantial financial resources and commanded a significant amount of labor. Decades later, in a time of rising populism, Harrison would reinvent himself as a commoner in order to win the presidency, portraying himself as a man who lived in a log cabin and drank seedling hard cider just as ordinary Americans did. But in 1801, it was grafted fruit trees, not wild seedlings, that helped him demonstrate his gentility and rise in the world.

There were other regions in Ohio that had little need for Chapman's seedling trees in 1801. To the west of the Virginia Military District, Cincinnati was emerging as a burgeoning urban center with a commercial farming hinterland. The farms nearest the city were cleared and fenced by 1800 and possessed small orchards of apple and peach trees. By 1803, farmers could purchase grafted nursery stock of Spitzenburgs, Pound Sweets, Pippins, and other popular commercial varieties of apples.[34]

The poverty of the Chapman and Williams families and the plenty of the Worthingtons were two extremes of the Ohio immigrant experience. In between were the experiences of families like the Staddens from Northumberland County, Pennsylvania. The Staddens were among the first pioneers to arrive in central Ohio's Licking Valley. Isaac traveled alone to the region in the winter of 1801, established a claim, built a primitive log cabin, cleared a small plot, and set out some corn. In the late spring he returned to his family home in Pennsylvania, loaded his wife Catherine, their two children, and their most important household goods into a wagon, and headed west. The Staddens possessed enough wealth to bring a plough and the basic tools a farmer needed. There was no fancy pier glass in their solitary wagon, nor could they carry hundreds of expensive and fragile grafted trees. Just before they left their old home Catherine Stadden carefully dug up three small suckers from her favorite dooryard apple tree and placed them gently atop the other items in her personal chest. They headed west across Pennsylvania, crossed the Ohio River at Wheeling, and took Zane's Trace down to Fairfield County before turning north and blazing their way through the forest to Isaac's cabin in the Licking Valley. Not long after they arrived, Catherine Stadden planted the precious root sprouts in the yard of her new home.[35]

John Chapman appeared in the Licking Valley with his packhorse laden with leather bags filled with apple seeds that very same year. His appearance in 1801 might not have been the result of random wanderings, as Cooley family tradition recalls that Asahel briefly moved his family into the Licking Valley about that time but returned to Ohio Company lands near the Ohio River after they contracted fever and ague, conditions he blamed on the poor drainage of the land and the absence of springs.[36] With only a handful of scattered cabins and still no roads, the Licking Valley looked to John like a place that might need his trees. More settlers were arriving, some poling and pulling their worldly possessions the thirty miles up Licking Creek from its confluence with the Muskingum at Zanesville, others forcing their wagons north from the Trace across open land, just as the Staddens had done. About a dozen families who trace their roots in the Licking Valley back to the early nineteenth century proudly claim that the first fruit trees on their lands came from John Chapman's nurseries.[37]

The Staddens were not among them. When an 1863 county history claimed that John Chapman was responsible for the first apple trees in the Licking Valley, that he had partnered with Isaac Stadden in this venture, and that "nearly all of the orchards of Licking County were planted" from John's seedlings, Catherine Stadden, by then a widow, was not pleased.[38] In her later years Catherine was proud to be a member of the Licking Pioneer Society, and she was appalled to hear that her late husband had been connected to the ragged wanderer Johnny Chapman. It is true, she asserted that her husband had a partner when he set out the county's first orchard, but that was not "Johnny Chapman" as she called him. Stadden's partner was a man named Johnny Goldthwait. The orchard that Mr. Stadden and Mr. Goldthwait had set out was filled with grafted trees of the best commercial varieties, not the shabby seedling trees that Chapman planted. "Johnny" Chapman was in the Licking Valley in those early years, she acknowledged, but she dismissed the idea that he played any significant role in the development of the valley's abundant orchards. "'Johnny' was not a practical man," Catherine Stadden insisted, and she believed he started only one seedling nursery in the entire county. "It was neglected, left unenclosed, so that domestic animals browsed upon it, and it afforded but few, if any, trees for transplanting."[39] In contrast, Johnny Goldthwait was "an experienced horticulturist" who devoted great care and attention to his nurseries of improved, grafted varieties.

Catherine Stadden's dissent has largely been ignored or dismissed by the tellers of the Johnny Appleseed legend in later days, some of whom imply that she may have been jealous of Chapman's fame.[40] But the story of Johnny Chapman and Johnny Goldthwait in the Licking Valley reveals the ways seedling and grafted apples came to be symbols of a real cultural divide among Ohio's early emigrants. Contemporary observers of the frontier often commented on the different types of people who occupied it. Eastern elites like Yale College president Timothy Dwight and Philadelphia physician Benjamin Rush celebrated the transformation of the interior "from forest to fruitful field," but they did not write in flattering terms of all of the people who participated in it.

Rush explained this transformation as occurring in three overlapping phases.

Western lands were first a place of "retreat for rude and even abandoned characters" seeking isolation from civilized society. This group did not linger long but nonetheless performed a useful service by starting the hard labor of clearing forest.[41] These border ruffians were soon replaced by very poor farmers of limited resources, who established crude farms characterized by corn planted amid the trunks of girdled trees, hogs let loose to forage in the forest, and disorderly orchards of wild peach and apple trees. The tenure of the second-stage settlers was almost as brief as the first, Rush believed, because they practiced such a primitive and improvident form of agriculture that they were unable to make a sufficient living, especially as men with more ambition, capital, and industry appeared. These third-stage settlers, interested in profits, set about making permanent, sustainable improvements. They cleared the fields of stumps and rocks and plowed it deeply. They manured the soil; improved the neglected orchards through top-grafting, pruning, weeding, and fencing; or set out new orchards from purchased grafted stock. They built mills and larger barns and thereby completed the transition from forest to fruitful field.[42] Timothy Dwight added to this progression narrative a regional conceit, insisting that the final stage of transformation was most effectively carried out by settlers from New England, a judgment the Worthingtons and Tiffins of Ohio would have contested.[43]

This notion of different frontier cultures has filtered down to the present day in the form of a debate about the connection between the pioneer settlers and the market economy. Some scholars argue that pioneers sought out the frontier to escape the uncertainty of market economies, and to secure personal independence through a self-provisioning lifestyle; others insist that all men are natural capitalists and eagerly embraced the opportunity to improve their economic standing by engaging in market behaviors as soon as the market was available to them. If pioneers were producing primarily for home consumption, it was only because inadequate transportation methods denied them access to markets. A balanced assessment suggests that the values of independence and improvement existed side by side on the frontier, while different types of emigrants prioritized one over the other. One historian of the trans-Appalachian frontier divided its emigrants into two broad categories: "incipient capitalists," who were eager to transform their new homes into settled, civilized communities connected to wider markets, and "transients," who appeared to relish the isolation of the new lands, embraced an independent lifestyle focused on self-sufficiency rather than market production, and often moved west as soon as increased populations and increased land prices made that lifestyle no longer viable.[44]

Catherine Stadden understood that apples were an indicator of the class of pioneer to which a family belonged. An untidy orchard of seedling trees suggested a pioneer had little ambition beyond subsistence; a well-kept orchard of grafted trees indicated that a family had ambitions to produce for wider markets. She was happy to have her husband associated with John Goldthwait, but not John Chapman.

Curiously, John Chapman and John Goldthwait shared a common Connecticut Valley origin. Both were raised in the Springfield area. The Goldthwaits owned a half share in the Ohio Company and were among those who gathered at Longmeadow in 1788 before setting off for Marietta. "Peculiar yankees" was a phrase frequently used to describe the Goldthwaits, a family of schoolteachers known in the Springfield area for their bookishness. John Goldthwait followed his father and became a schoolteacher as well, first teaching near Athens on Ohio Company lands but then moving his family farther into central Ohio. Goldthwait was a pious man, a devoted and enthusiastic Methodist. The grafted stock he brought with him to the Licking Valley he acquired at Israel Putnam's nursery along the Muskingum River.[45] After helping Isaac Stadden set out the first orchard of grafted stock in the Licking Valley, Goldthwait found a teaching position about twenty miles to south and west, near the village of New Salem, not far from the burgeoning town of Lancaster. There, he established a nursery of grafted apple trees and was soon supplementing his teaching income by selling trees to the growing population of immigrants tumbling off the Zane Trace.[46]

Goldthwait's passion for grafted apple trees became the basis of a joke that circulated for decades in Fairfield County. The joke recalled the day when two Lancaster lawyers traveled out to the country to see Goldthwait's nursery. The lawyers happened to be partisan Jeffersonian Democratic Republicans, while Goldthwait was devoted to the rival Federalist party. After leading the lawyers past neat rows of his Pippins, Russets, and Seek-No-Furthers, Goldthwait showed off a tree he called the Federal apple. One of the lawyers responded "you have shown us your Federal apples. Now show us your Democratic ones." Goldthwait led the two lawyers to the edge of the field, where a small scrubby volunteer seedling "with a few knotty apples on" had sprung up. "'That,' he sneered, 'Is the Democratic apple.'"[47]

The story, no doubt repeated frequently by local Federalists (and later

Whigs) in the early years of Fairfield County, had layers of meaning that might escape the modern hearer of the tale. It was not the kind of joke that worked when the parties were reversed. For improvement-minded Federalists, the seedling tree had become a symbol of the idle and lazy subsistence farmer, in their mind the core constituency of a Jeffersonian Republican party that idealized the simple yeoman farmer. Other contemporaries made similar political associations with wild apple trees. A few years later, down in Rome Township, Lawrence County, along the Ohio River, Zebulon Gillette discovered a sucker growing below the graft of a tree he had purchased from Israel Putnam's nursery upriver at Marietta. He cut the sucker off and handed it to his son Joel, telling him "this one's a Democrat, you can have it." Perhaps Zebulon should not have been so disdainful of the wild sucker. Joel rooted it and planted it. The tree thrived and eventually bore beautiful deep red round apples, and the Rome Beauty variety was born.[48]

As cultural symbols, the grafted and seedling apple tree were crude stereotypes. But they nonetheless had resonance in the early nineteenth century. Seedling apple trees at various times were associated with the poor, the idle, the unambitious, the primitive frontier family eking out a subsistence, those who sought isolation from markets, the common southerner, and the cider-drinker. Grafted apple trees were associated with the man of capital, the industrious, the striver, the gentleman farmer, the improver, the Yankee, and the teetotaler. As influential as these stereotypes were, real settlers often defied them. Democrat and Southerner Thomas Worthington emulated Jefferson in establishing orchards of grafted fruit at Adena. Yankee John Chapman maintained a lifelong attachment to the seedling tree. Furthermore, personal circumstances often dictated the kinds of trees a family planted, and the choice of seedling stock was not immutable. A poor but ambitious family might set out a seedling orchard with the aspiration to top-graft these trees with marketable varieties at some later date. But even after fruit grafts became readily available, the ungrafted seedling orchard persisted in Ohio for decades. Some Ohioans would be forever content with their seedling orchards.

John Chapman earned his reputation as a purveyor of trees that sprang from *seeds*, rather than from a nursery of transplanted grafted stock. It was not "Johnny Apple*tree*" that he became, but Johnny Apple*seed*. This distinction is lost on most modern hearers of the tale, but it had real significance in John's own time. In later decades of the nineteenth century, when ungrafted seedling orchards had disappeared, some Ohio children might come to imagine that an apple tree came from a commercial nursery in the same way that a child today might believe that an apple comes from Walmart. But understanding the cultural meaning of seedling and grafted apple trees in the early nineteenth century is critical to understanding who John Chapman was and how he was received in the communities in which he lived. John Chapman provided seedling trees to the poor, to the "suckers," to the "Democrats" of Ohio; John Goldthwait sought a different kind of customer.

OWL CREEK AND MOHEKAN JOHN'S CREEK

From the Licking Valley, John Goldthwait had moved south toward the Zane Trace and burgeoning market towns. John Chapman went north, further into the wilderness, to the site of an old Indian village on the banks of another stream that fed into the Muskingum River watershed. Today it is called the Kokosing, but the first white settlers knew it as Owl Creek. Chapman may have first explored the area on the same 1801 trip that brought him to the Licking Valley, but probably he arrived no later than 1803. By 1806 he had seedling trees for sale to new settlers streaming into the region.[49] While Owl Creek Valley was a mere twenty-five miles from the Licking Valley lands settled by the Staddens, in 1801 it was a degree more isolated. While Native Americans occasionally hunted and camped in the Licking Valley, the Owl Creek region was still more Indian that white. But that would change quickly. One of John's first nurseries in the region, on the "Indian fields" near a recently abandoned Delaware village, was just ten miles below the Greenville Treaty Line, which until 1805 still served as the formal boundary between Indian and white settlement.[50] The boundary, however, was permeable. Indians still hunted and camped below the line, and some aggressive whites were already staking claims above it, in anticipation of further treaty concessions William Henry Harrison would press on the local tribes in 1805, when he confined their settlement to a few small polyglot villages. Still, the local Indians retained rights to hunt and fish on surrounding lands, and trade between whites and Indians meant that their interactions were frequent.[51]

By that same year, white boosters had platted the town of Mt. Vernon on the north side of Owl Creek and were selling off town lots to new arrivals.[52] Every year more migrants moved into the region, sometimes settling on land before government surveyors had the opportunity to mark it off. Still, there were no roads to the region, so access was limited to a few Indian paths or by a 150-mile-long water route from Marietta up the Muskingum River and several tributary streams. Owl Creek flowed fast after spring rains, but it could be just ankle deep

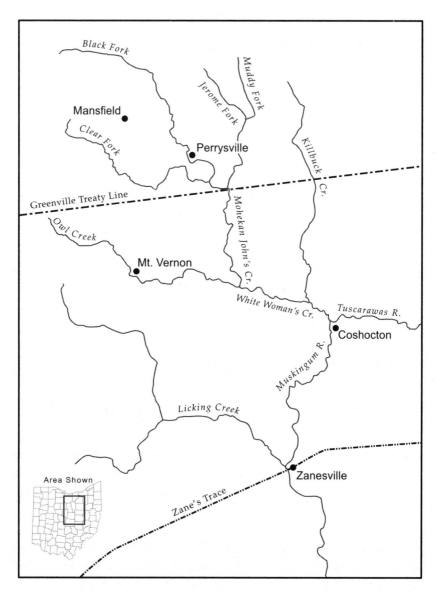

Johnny Appleseed country, ca. 1810. Claudia Walters, University of Michigan–Dearborn.

in late summer. As they neared the site of Mt. Vernon, travelers would have to pole and pull up the creek in nothing larger than a pirogue.

The first white migrants to Owl Creek were some of the roughest of frontier characters. Among the most notorious of these was Andy Craig, who "was living there in a little log hut, with a great raw-boned woman as his wife." Craig and his partner were fugitives of sorts, as the unnamed woman was legally married to another man and residing in Wheeling, Virginia, when she took up with Craig and the two headed into the interior to find blissful isolation. But Craig and his partner were not alone for very long. Some of those who followed were improvers eager to profit from the wave of migrants who followed; most were families of very modest means, hoping to secure a land claim and a competency for themselves and their families.[53]

In 1806 John Chapman was selling seedling trees to newcomers in Mt. Vernon, and fifteen men, Chapman among them, gathered to vote in the new county's first election. Three years later, Chapman became a landowner, purchasing two "town lots" in the village, both of them several blocks below the center of settlement and abutting Owl Creek. That it was eight years between the time of John Chapman's first recorded trip into Ohio and this first modest acquisition of land was not unusual. In 1810, still less than half of the nearly 35,000 males residing in the state who were twenty-six years or older had managed to acquire land. Absentee speculators dominated the land market, controlling about half of the land sold. Large scale speculators—the top 1 percent of landholders in Ohio, controlled about one-quarter of all the state's lands. Opportunities for poor migrants there were certainly far better than they had been in northwestern Pennsylvania, but about half of Ohio's immigrants remained tenants or squatters in 1810.[54] Despite his new landed status, there is no evidence that he ever established a residence on these lots, and one appears to have been erased when the river subtly altered its course. In all likelihood, he made the purchase to protect his claim to another creekside seedling nursery he had established. In 1810, Chapman remained primarily a squatter, setting out his nurseries on land he did not own. But Ohio land deeds required the purchaser to state a county of residence, and in 1809 John Chapman listed his as Knox County.[55]

Despite these signs of respectability, Chapman was not settling down. He was still constantly on the move, traveling back to Pennsylvania, no doubt also going down to see his family in Marietta, and planting more apple seed nurseries. Even as he maintained his presence in Mt. Vernon, he was actively setting out new seedling orchards on lands along the many creeks and streams of the upper Muskingum watershed. East of Mt. Vernon, Owl Creek flowed through a valley of low hills, interspersed with significant stretches of river plains, some only recently abandoned by Indian farmers. Owl Creek eventually joined the south-flowing waters of Mohekan John's Creek to form the wider, more constant waters of White Woman's Creek. Much of John's seedling nursery business in the first decade of the nineteenth century was centered near the confluence of these rivers. He reportedly had two or three nurseries on the lower reaches of Mohekan John's Creek, waters that flowed through a rugged gorge, uninviting for settlement. But Chapman's Mohekan nurseries—two near the banks of the river and one on a midstream island—were just a few miles away from flatter, more tillable lands, making them ideal for his purposes. The imposing geography of the gorge deterred farmers from making claims to the land where John planted his squatter nurseries, but they were close enough to more attractive farmland to be easily accessible to his customers.[56]

It was in these central Ohio settlements in the upper reaches of the Muskingum watershed that John found a home of sorts among a group of mostly Scots-Irish Pennsylvanians and upland southerners, with a scattering of Yankees added to the mix. One resident of the region recounted Chapman's strategy a few decades later. "He would clear a few rods of ground in some open part of the forest, girdle the trees standing upon it, surround it with a brush fence, and plant his apple seed. This done, he would go off some twenty miles or so select another favorable spot, and again go through the same operation. In this way ... he rambled from place to place, and employed his time, I may say his life. When the settlers began to flock in, and open their 'clearings,' old Appleseed was ready for them with his young trees; and it was not his fault if every one of them had not an orchard planted out and growing without delay."[57] Yet Chapman's practices, which depended on only the labor he himself could provide, resulted in vulnerable nurseries with high attrition rates. Another early settler who recalled these nurseries suggested that the primitive log and brush fences, and the fact that they were rarely tilled and weeded, did not bring great yields of trees. "This protection was not sufficient to keep out the deer, who browsed off the young trees; and sometimes in the fall the fire would run through it, and in consequence the nursery did not flourish—But [some] trees were taken from [these nurseries] and set out at different places, which have borne and are still bearing fruit."[58]

Fire and browsing deer were not the only threats to Chapman's nurseries. The fact that he planted them on land he did not own meant that when he returned for them, he could find the land occupied by a new settler, who might not be willing to grant him access. Probably for this reason, midstream islands, and

John Chapman preparing a seedling nursery. Chapman planted most of his early nurseries on land he did not own, alongside creeks and rivers to provide easy access to water, about twenty miles apart in territory just being settled. Many were lost to floods, fire, browsing animals, or landowner claims. *Harper's New Monthly Magazine*, Nov. 1871, 830.

small patches of ground surrounded by otherwise inhospitable terrain appear to have been some of his favored locations for planting. One of his lower Mohekan nurseries was reputedly on land owned by an Eastern speculator who had no intention of moving to Ohio to settle the land but instead hoped to make a windfall profit by reselling the land after the region was densely populated and land values had increased. The two Mt. Vernon town lots are the only recorded land purchase by John Chapman in the first decade of the nineteenth century.[59]

John's method of business seems to the modern hearer astoundingly casual. As he was constantly on the move, he generally made arrangements with a nearby family to oversee the sale of his trees. When he encountered a customer nearby, he might write a note to the caretaker, like one preserved by the Rice family, which simply read "Mr. Martin Mason, Sir, please let Eben Rice or bearer have thirty-eight apple trees and you will oblige your friend, John Chapman, Richland Co., Ohio, August 21, 1818." We cannot be sure what debt the thirty-eight trees were intended to settle, but they may have been to cover the cost of John's board, as he was a frequent overnight guest of the Rice family. That the note included the language "or bearer" revealed that informal notes like this one were transferable, and often stood in for currency in a cash-scarce world.[60] Much business was carried out through barter, or locals with an established relationship to the nearest merchant had their debits and credits recorded in a simple store ledger. Local banks, many of them unlicensed by the state, issued paper in denominations as low as twelve and a half cents. And "sharp shins," twenty-five- and fifty-cent coins cut into five or six pie wedges, were also used for small purchases. But a handful of sharp shins could wear a hole through a tow linen pocket as they jostled around in a single day's travel.[61]

This is how Benjamin Rush's suckers lived on the Ohio frontier. And given the meager resources the first settlers of the region had, it is no surprise that they chose Chapman's seedlings, which he generally sold for three to five cents apiece, when they got around to setting out an orchard. Grafted stock, could it be had at all, might cost a farmer upward of twenty-five cents a tree in these early years.[62] And Chapman appeared to be willing to negotiate on price. For the poor, he would accept whatever they could pay, in cash or in kind; he accepted promissory notes as well, and by most accounts he was remarkably casual about collecting on debts owed and did not trouble pursuing those who did not pay. Such behavior is viewed today as extraordinarily charitable, and it was sometimes recognized as such in its own time. But it was not completely out of line with late-eighteenth- and early-nineteenth-century practices of exchange on the frontier, where debits and credits were listed in store ledger books, but ledgers were rarely balanced. That a certain amount of debt would never be repaid was part of the practical reality of frontier economies.

John's casual indifference to payment and debts was logical in the context of the product he was peddling. The seeds he planted were free, and the labor he invested in planting them quite minimal. Even a small creekside nursery could produce a large volume of trees, and he no doubt soon found himself with a large surplus of trees. One story about John's largest nursery on Mohekan John's Creek claimed it covered five acres of land and that he sowed sixteen bushels of apple seeds on it. A previous biographer playfully estimated that such a nursery might have yielded over four million trees after a 20 percent loss was deducted to account for theft by hungry chipmunks and other varmints.[63] That the size of the nursery and the number of seeds sown were greatly exaggerated doesn't undermine the larger point. Even a small sack of apple seeds could potentially produce more than enough trees to meet the needs of every farmer within a day's walk. But the seedlings from any particular nursery also had a limited "shelf life." Unsold trees would eventually grow too big for transplanting, so accepting a small sum, some goods, or a promissory note for them was better than receiving nothing.

It was not simply price and availability that made Chapman's seedling trees desirable for his poor neighbors. In surveying the history of agriculture, Noel Kingsbury noted that precapitalist traditional farmers pursue strategies that favor reliability and food security over yield, and there is no better protector of food security than genetic diversity. An orchard of ungrafted seedling apple trees increases the likelihood that some trees will be resistant to whatever climactic or natural stresses arise in any given year. An orchard of grafted trees, lacking that natural diversity, will in good years dramatically outproduce the seedling orchard, but it is also more vulnerable to catastrophic losses as the result of a late frost or the appearance of some new pest.[64] Furthermore, Chapman's strategy of planting his nurseries every twenty miles or so allowed farmers to select those seedlings that were already proving their ability to thrive in their particular microclimate. Importing expensive grafted trees from even Putnam's nurseries one hundred miles south at Marietta carried the risk that they might prove to be unsuitable for local soils and climate.

John Chapman was a poor man among poor men in the Mohican Valley. He had no fixed abode but instead hired his labor out in exchange for room and board and perhaps a little cash or goods. There was a shortage of labor in Ohio, and the going rate was fifty cents a day. His clothes were primitive and his feet bare, but in the early years of settlement in central Ohio, such attire didn't make him entirely stand out. "Ladies and gentlemen, *when they clothed their feet at all,* dressed them in moccasins," one early chronicler of the valley recalled. Men generally wore clothing made of buckskin and homemade flax linen, women's dress was generally of raw cotton and homemade linen. Calico was an expensive luxury few could afford. "An excellent and industrious girl, as late as 1822 or 1823, toiled faithfully six weeks for six yards of calico, which, in those primitive days . . . was deemed sufficient for a dress. The lady who appeared in the first calico dress, attracted, it may be supposed, considerable attention in 'the settlement,' and was regarded as an aristocrat."[65] John Chapman's ragged garments, which drew much attention in later years, probably elicited little commentary before 1820.

GREENTOWN

Even as more white settlers arrived in the region, the Native Americans who also considered central and northern Ohio their home remained. In 1805, in the Treaty of Fort Industry, the Wendat, Delaware, Munsee, Shawnee, Ojibwa, and Potawatomi, who had claims to a half million acres of land along Lake Erie, reluctantly surrendered them in exchange for annual payments. They retained their rights to hunt and fish in the region, but settlement was relegated to a few small reservations, including two small villages along tributaries of Mohekan John's Creek: Greentown on the Clear Fork and Jerometown on Jerome's Fork. It did not take long for white settlers to begin swarming into the newly opened lands, and the Indians at Greentown and Jerometown found themselves surrounded by white settlers and the signs of their presence. John Chapman was among them, getting out ahead of the encroaching tide, planting his nurseries on the many tributary streams that stretched deep into a vast area to the north into recently surrendered Indian lands. Mohekan John's Creek received its water from tributaries flowing across more gently rolling lands to the north and the west. Today they are known as the Clear, Rocky, Black, Lake, Jerome, and Muddy Forks of the Mohican River, and John may have had nurseries along every one of them.

By 1809, John had at least one nursery on the Clear Fork on lands north of Greentown, and he was among a small group of men who prepared a celebratory meal on the newly platted town square of Mansfield, not far from the banks of the Rocky Fork. He also probably had nurseries along the Lake Fork to the east by this time. By 1812, possibly earlier, John had crossed the low divide that separates south-flowing waters from north-flowing ones and was living and planting trees along the Huron and Vermilion Rivers, which flowed into Lake Erie.[66]

In legend, Johnny Appleseed is frequently described as a great friend to Indian peoples, and it is commonly asserted that they innately trusted and welcomed him. But the claims of Chapman's close relationships with local Native Americans are vague, and the handful of specific stories that survive about Chapman's encounters with Indians do not support the legend. One story recounts that Indians near Greentown stole about twenty ponies from Chapman but that he made no effort to retrieve them, concluding that the Indians needed them more than he did. And the few stories about Chapman and Indians that survive from northwestern Pennsylvania, mentioned in the previous chapter, recall his efforts to escape Indians who were out to harm him.[67]

The best we can say about Indian-white relations in central Ohio during these years was that they were generally friendly, except when they weren't. Periods of mutually beneficial and mutually welcomed trade were punctuated by acts of extreme brutality, carried out in vengeance by whites or Indians whose memories of unanswered atrocities were still fresh. These bloody incidents assured that relationships between whites and Indians never became truly relaxed and easy. Subtler forms of aggression were more common—horse stealing by Indians, movement onto Indian lands by whites. And there were the unresolved disputes that festered: whites who resented Indian ponies invading their corn fields and destroying their crops, Indians who felt cheated in their dealings with local white merchants. Even in the best of circumstances, mutual distrusts and resentments simmered. But for the most part, neither whites nor Indians sharing central Ohio lands in the first decade of the nineteenth century desired open conflict. The Indians of Greentown and Jerometown were under the leadership of old chiefs who understood too deeply the loss and pain of war and encouraged their people to live peacefully with encroaching whites.

Just a few days' walk west, Indians living in a region still economically tied to British trade were more inclined to resist the American government's insatiable demands for more land and to chafe against the patronizing tone of American missionaries who told them that they would be happier if they simply learned to live like white people. Presidents Jefferson and Madison pursued an Indian policy that encouraged Indians to adopt white ways of living, if for no other reason than it required less territory and would therefore open up more lands for white settlement. William Henry Harrison, territorial governor of Indiana, served as "Mr. Jefferson's Hammer" in pressing this policy on Indians.[68]

The native resistance movement found inspiration in the visions of a reformed Shawnee drunk named Lalawethika—the Noisemaker. The story of his emergence as the Shawnee prophet paralleled that of the Seneca prophet Handsome Lake, who had been drawing attention to the village of Burnt House in the days when Chapman was still in Pennsylvania. Once a troubled alcoholic, Lalawethika awakened from a deathlike trance with visions of another world, a message from the spirits, and a new name, Tenskwatawa—the Open Door. He swore off alcohol and the white man's ways and encouraged others to do so as well.

Tenskwatawa's visions captured northwestern Indians' anxieties perfectly, as they confronted a crisis that was not simply cultural and spiritual but environmental as well. As white settlers had moved in and replaced forest with field and orchard and killed predators to make their new homes safe for livestock, the very land itself underwent dramatic change, and not simply in the obvious ways brought about by the settlers' axes and rifles. The very flora and fauna of Ohio began to undergo a disconcerting transformation. European honeybees—"the white man's flies"—spread into the woods in advance of white settlers. Grazing livestock carried the seeds of Old World grasses west in their bowels, and indigenous ground cover soon gave way to a carpet of alien invaders. Indians dubbed the broadleaf weed today known as the plantain "Englishman's foot," as it appeared to spring from the ground wherever white men trod.[69] In one of Tenskwatawa's visions white settlers appeared on Indian lands as a great crab with mud dripping from its claws, just emerged from the sea. "Behold this crab," he told his Indian followers, "it comes from Boston, and it brings with it part of the land in that vicinity." But Tenskwatawa promised his followers a way to defeat this monster. Were they to abandon white ways and return to their native traditions, he promised, "I will overturn the land, so that all the white people will be covered and you alone shall inhabit the land."[70]

The prophet's message spread rapidly across the Michigan and Indiana territories and parts further west. Indians along the Maumee River in northwest Ohio were also attracted to the message. The Indians of Greentown and Jerometown certainly shared the anxiety of their kin to the west, but most did not join the prophet's movement. Weary of war, surrounded by and increasingly dependent on their white neighbors, the Greentown and Jerometown Indians continued to pursue a path of accommodation and assimilation. Indeed, they played a role in the environmental transformation occurring around them. When the Ohio government began offering bounties for the scalps of the wolves that preyed on white livestock, it was Indian hunters who did much of the killing.⁷¹ Greentown Indians killed wolves, and Chapman planted apple seedlings. Both served the great crab from Boston.

It is tempting to want to see John Chapman as a mediator between white and Indian worlds, a voice for peace in a world turning toward war. But ultimately the vision of Tenskwatawa and that of John Chapman were at odds. Fortescue Cuming, who crossed Ohio in these years, captured the perspective of most white settlers when he wrote in his diary of the demoralizing sight "of ridges beyond ridges covered with forests, to the most distant horizon." This view for Cuming was "dreary and cheerless, excepting to a mind which anticipates the great change which the astonishingly rapid settlement of this country will cause in the face of nature in a few revolving years." In his mind's eye, he substituted for the unbroken forest a vision of "valleys divested of their trees; and instead of somber forest . . . the rivers and brooks, no longer concealed by the woods, meander through [these valleys] in every direction in silver curves, resplendent with the rays of a golden sun, darting through an unclouded atmosphere; while the frequent comfortable and tasty farm house,—the mills—the villages, and the towns marked by their smoke and distant spires, will cause the traveler to ask himself with astonishment, 'So short a time since, could this have been an uninhabited wilderness?'"⁷²

There should be no doubt which vision the wandering apple tree planter shared. Imagining Johnny Appleseed as a person who fit harmoniously in both the Indian and white worlds has allowed white Americans of later ages to imagine an alternative version of the transformation of American forest to field. But Chapman was an eager agent in this transformation.

Under the guidance of Tenskwatawa's brother Tecumseh, this pan-Indian spiritual revival soon evolved into an alliance of Indian peoples determined to resist further white encroachments onto their lands. While Tenskwatawa encouraged Indians to reject assimilation and return to traditional ways, his brother Tecumseh encouraged them to see that the practice of accommodation in the decade after the defeat at Fallen Timbers had failed and needed to be replaced with a policy of resistance.

As Tecumseh and his followers became increasingly resistant to William Henry Harrison's demands for more land concessions, incidents of white-Indian violence escalated, finally exploding into frontier war at Tippecanoe Creek in Indiana in 1811. A setback for Tecumseh and his followers, it did not reverse their determined mood of resistance, and white westerners, eager to find an external source behind the new Indian assertiveness, blamed the British and joined with Eastern politicians in a rising chorus for war against the former mother country. Most white Ohioans embraced the call for war enthusiastically and were confident that the young United States would easily destroy British influence in the West, settle the Indian problem once and for all, and annex British Canada.[73]

When war came, Madison settled on a strategy of invading Canada at three different points: Quebec, Niagara, and Detroit. Michigan's territorial governor, William Hull, was put in charge of planning and executing an invasion from Detroit. On May 25, Hull gathered an army of two thousand in central Ohio, mostly Ohio and Kentucky volunteer militia. Hull stoked the already rising fire of anti-Indian feeling by delivering a speech to his soldiers warning them that they were about to march into "a wilderness memorable for savage barbarity."[74] In June, Hull's troops began a slow march north, establishing forts along the way, not arriving in Detroit until early July, where he crossed into Canada at the tiny village of Sandwich and began setting up defenses. Hull was an alcoholic and a bumbling military leader who inspired little confidence in his officers and men. His behavior was erratic and counterproductive. He presented himself to the residents of upper Canada as their liberator, then granted permission to his troops to plunder them. The soldiers did not hesitate, and looting and destruction ensued. The fences and orchards of Sandwich were ripped up and used for firewood. When, in late July, Fort Michilimackinac in northern Michigan fell to a small force of Canadians and Indians, Hull panicked as he imagined swarms of savage Indians descending upon him from the north, and he ordered his troops back across the river. His junior officers considered mutiny, and morale among the American force collapsed. A few weeks later, when the British and their Indian allies crossed the Detroit River and paraded in front of Hull's fortress, the commander quickly offered a complete surrender without consulting any of his officers.[75]

It was a humiliating capitulation and resulted in a dramatic change in mood among white Americans in the region. The intense hatred for "savage" and "barbarous" Indians that Hull had promoted among his troops had been paired with a supreme confidence in victory. That confidence turned to hysterical fear of Indian savages as news of Hull's surrender spread across Ohio. The terms of surrender allowed for the parole of the 1,600 Ohio and Kentucky volunteers, sent home to sit out the war, yet still filled with anger, fear, and humiliation. This certainly presented a precarious new reality for those Indians in central Ohio still hoping to stay out of this war, as the news of Hull's surrender struck terror into the hearts of Ohio settlers, who in times of conflict were not inclined to make distinctions between peaceful and belligerent Indians. Captain Pipe, the chief at the small village of Jerometown, recognized the danger his people were in and did not wait for instruction. He gathered them and trekked to Cleveland, where they professed their loyalty to the United States government and placed themselves under the "protection" of American officials. The Indians at Greentown chose to sit tight.[76]

Despite the fact that the Indians at Greentown had continually accommodated American demands for more land, and had resisted Tenskwatawa's message, white leaders in central Ohio now perceived them as potential enemies in their midst and decided to remove them to a place where they could be monitored and controlled. The local militia commander pressured the Reverend James Copus to accompany troops to Greentown because he knew the minister was on very good terms with the Indians and might help persuade them to leave peacefully. Copus reluctantly complied, assuring the Greentown Indians that their removal was only temporary and if they cooperated no harm would come to them or their village. The Greentown Indians had little choice but to comply. Experience told them how vulnerable neutral Indians were in time of war. But they were not so naïve as to be surprised when they turned to look back at their village and saw smoke rising as a few militia stayed behind to torch it. The militia had cynically exploited James Copus's relationship with the Indians and turned him into a liar in their eyes, although in all probability Copus knew nothing of their sinister plan. The militia first took their prisoners to Mansfield and held them under guard in a ravine south of the town square. When a Sandusky Wendat named Toby, who had come to Greentown to retrieve his daughter, tried to slip away from camp, the militia pursued them and shot Toby, who fell wounded to the ground. His daughter escaped. Toby begged for mercy and professed his commitment to peace, but a militia member ignored these pleas and drove his hatchet into Toby's skull. Others made a half-hearted attempt to bury his body, but according to local tradition, for several years Toby's ribs rose out of the soil at the place where he fell. Perhaps no action did more to insure the actualization of white fears and to drive some of Greentown's young warriors to violence than the destruction of their village and the murder of Toby.[77]

In the midst of these events, the thirty-seven-year-old John Chapman was traveling between his Mohekan nurseries and lands along the Huron River to the north, right in the middle of the conflict. He had already developed a reputation for extraordinary meekness, and he was not among the 1,600 Ohio and Kentucky men who volunteered for war. But if he did not share the Indian hate of his white neighbors, in the wake of Hull's capitulation, he certainly shared their fear. That summer he was spending much of his time at the home of Caleb

Palmer, one of the first settlers along the Huron River. In the case of a British-Indian invasion by water, the sparse settlements along the Huron, not far from the lake, would be exceptionally vulnerable. In such an environment, it did not take much to set off an alarm, and according to a local story, when Chapman and Palmer heard a few shots ring out from the direction of another settler's cabin, Chapman took his rifle and set out to investigate. The shots turned out to be nothing more than the neighbor's successful efforts to take down a deer. Chapman stopped to help the neighbor skin and dress the deer, and when that was all done the neighbor gave him a chunk of venison for his trouble. By the time Chapman returned, Palmer was in a panic, and mistaking John for an Indian, almost shot him.[78]

The story contradicts the later versions of the Appleseed myth that suggest he never carried a gun and would not eat meat or harm an animal. If Chapman eventually did reject firearms and embrace vegetarianism, he did so at a later date. The account also calls into question the more exaggerated accounts of his pacifism. Chapman responded to the gunshots by grabbing his own gun before going to investigate.

Not long after the deer incident, a more thorough panic spread throughout the region, as rumors spread that a force of Canadians and Indians had landed at the mouth of the Huron on Lake Erie. The report proved to be false—the boats landing were in fact filled with paroled, defeated volunteers from Hull's army. But according to local tradition, John Chapman was responsible for spreading the panicked rumor, which caused the families of the Huron Valley to flee south through heavy August downpours to settlements at Mansfield and Fredericktown.[79]

While this alarm turned out to be false, the phantom menace eventually turned into a real one, as Indians seeking vengeance for the destruction of Greentown and murder of Toby struck back. They shot and scalped a Mansfield merchant named Levi Jones on the banks of the Rocky Fork, a short distance from one of Chapman's nurseries. Expecting that more was to come, families gathered in the blockhouse in Mansfield, as rumors of additional murders exploded. The families in the blockhouse felt quite vulnerable—the local militia were still away in Piqua guarding the imprisoned Greentown Indians. Someone needed to go for help, and John Chapman was charged with heading to Mt. Vernon, thirty miles south, to warn other settlers and get help. Chapman's race through the night from Mansfield to Mt. Vernon, warning settlers is among the most popular and frequently retold Johnny Appleseed tales. In most versions, Chapman ran the thirty miles barefoot, but in the first version to make it into a printed history, he was more sensibly and efficiently riding a horse. In the memories of some families who heard the words of this latter-day Revere echoing through the forests in the September night air, his words were biblically Jeremiac. "Flee for your lives. The British and Indians are coming upon you, and destruction followeth in their footsteps!" recalled Amariah Watson. A later version was imbued with even more religious imagery: "The spirit of the Lord is upon me," Chapman shouted, "and he hath anointed me to blow the trumpet in the wilderness, and sound an alarm in the forest; for behold, the tribes of the heathen are round about your doors, and a devouring flame followeth after them."[80] White fears of a British and Indian invasion of central Ohio proved to be unfounded. The British had not landed in Ohio. The murder of Trader Levi Jones was not the first blow in a planned campaign of retaliation against all whites. Jones's killers held a specific grievance against their victim related to a trade deal.[81]

Despite their fears, many white families were anxious to leave the overcrowded, hastily erected blockhouses and return to their farms. It was harvest season, the worst possible time for any Ohio people—Indian or white—to be forced from their homes. Stored corn meant subsistence and survival through the next year. The Greentown Indians marched off to camps in Piqua and the white families driven by fear to the protection of blockhouses shared a common anxiety that September. An attempt to retrieve some of that corn crop might have been what drew Indians back to the region, and it was that same driving to desire to save an annual harvest that tempted some settlers to leave the blockhouses and return to their farms. Among those who stayed put on their farms or returned to them in early September were a German settler named Philip Zimmer, his neighbor Martin Ruffner, and James Copus and his family. At the insistence of the local militia commander, nine militia were sent to Copus's home to protect him, an action that probably confirmed for Greentown Indians their suspicion that he had deliberately deceived them. On September 10, Indians killed and scalped Zimmer, his wife and child, and his neighbor Ruffner who had rushed to protect him. Again, the target was probably not random. Zimmer had drawn the Indian's ire in the past for allegedly tying clapboards or firebrands to the tails of Indian ponies that had invaded his cornfields.[82]

Five days later, a larger group of warriors, perhaps as many as thirty, attacked the militia sent to protect Copus family while they were bathing in a nearby spring, then surrounded the Copus homestead and poured fire into it for hours before finally retreating. James Copus and six militiamen died that day, in an event remembered locally as the Copus Massacre. At least two Indians

"THE TRIBES OF THE HEATHEN ARE ROUND ABOUT YOUR DOORS, AND A DEVOURING FLAME FOLLOWETH AFTER THEM."

John Chapman's barefoot night run from Mansfield to Mt. Vernon, Ohio, to warn white settlers of an impending Indian attack is among the most oft-told Appleseed legends. Stories frequently portrayed Chapman as a man equally at home in white and Indian worlds, but during the War of 1812 his loyalty to white settlers was unquestioned. *Harper's New Monthly Magazine,* Nov. 1871, 832.

also fell during the long battle.[83] Years later, a local historical society erected a monument to the white victims of the event. The monument also lists the name of John Chapman as somehow a participant in the events that September. But there is no specific evidence to support Chapman's presence there during the fighting.[84] He certainly knew the area well, and all of the white settlers along it. He allegedly had a nursery on Martin Ruffner's land.[85] Chapman's real role in these events is not preserved in credible sources. But in local mythology, Chapman was able to exploit the innate trust Indians placed in him to travel unharmed through this region and to alert whites to the presence of Indian belligerents in the area. These stories, of course, expose the contradictions and tensions within the Johnny Appleseed legend as it emerged in central Ohio in later decades. Was he a pacifist, a man who stood perfectly between the worlds of Indian and white Ohioans, innately trusted by both? The evidence seems to suggest that his fundamental allegiance was to his white neighbors. Was that allegiance limited to protecting white families from harm, or would he, as is recounted in the Caleb Palmer story, shoulder a rifle and use it if necessary when violence broke out? The collective evidence suggests that Chapman was a man of unusual meekness and gentleness in a rugged and violent frontier world. But ultimately he was allied with the white, not Indian, vision for the Midwest. He dreamed of a world of farmhouses, fields, and orchards.

A large-scale invasion of central Ohio by Indians and their British allies never materialized. But white settlers in the region continued to feel vulnerable, and many retreated with their families into southern Ohio until the danger passed. When Oliver Hazard Perry defeated the British Navy at Put-In Bay on Lake Erie in September 1813, and Tecumseh and his forces were defeated at the Battle of the Thames a month later, the threat passed. The war did not end until January 1815, but white Ohioans returned to the region in substantial numbers in the spring of 1814. Not just Ohio, but nearly all of the Old Northwest now seemed safe for white expansion. Indians never returned to Greentown or Jerometown—both villages were thoroughly destroyed by whites during the war. A few small Indian reservations in northern and western Ohio remained for decades, but by 1814 it was no longer in doubt that the region's future had been won for whites and white ways. This victory was achieved largely by plough and pruning hook; the musket and the rifle played only a minor role at the end of the drama. And it was by and large the suckers like John Chapman—the emigrating rural poor—who brought about this transformation.

But the world of Ohio's poor rural whites began to change again even before peace was signed. Contracts for war supplies had brought money into a region that had been cash-starved. More money and easier credit terms soon translated into a surge in land sales. Soldiers who first encountered the state while marching off to war saw much promise in the open lands of northern and western Ohio and returned with their families after peace came. An explosion of paper money produced by state banks, some licensed, some not, as well as an infusion of counterfeit notes, meant that after 1814, land sales boomed and prices rose. The once-isolated lands along Owl Creek, White Woman's Creek, and the forks of Mohekan John's Creek where John Chapman had set out his nurseries would soon experience another revolution.

CHAPTER FOUR

Walking Barefoot to Jerusalem

The War of 1812 settled the issue of whose vision for Ohio would prevail. Tenskwatawa's dream of a land that remained thick with forest and abundant in game, punctuated only periodically by creekside Indian villages and cornfields, quickly disappeared. Fortescue Cuming's dream of a land of farmhouses, fields, orchards, and market towns soon emerged. In the years after the war, change along the forks of Mohekan John's Creek came quickly. New settlers arrived every year. Trade with Europe was fully restored after a decade of embargos and disruption, and goods began moving more freely, but still with some difficulty, across rough roads, up and down seasonal creeks and rivers, over mountain passes, and down the wide Ohio and Mississippi Rivers to New Orleans and beyond. Bolts of machine-made calico fabric, some of it coming from as far away as Manchester, England, so coveted by the ladies of the Black Fork settlement, appeared with more regularity in the dry goods stores in Mansfield and Mt. Vernon. In the postwar years not just goods made long journeys; ideas did too.

MANCHESTER, ENGLAND

The distance from the banks of the Black Fork in central Ohio to Manchester, England, is about 3,600 miles in a straight line, much longer by boat and wagon. In 1817, the hamlet of Perrysville on the Black Fork consisted of a prairie cleared long before by its former Indian inhabitants, a handful of cabins, and some cornfields. The apple trees were too young to provide residents with a steady supply of cider, but wild honeybees, which had arrived before their human partners in colonization, were busy filling the cavities of rotting trees with an abundant supply of honey. In the thick forests that surrounded the Perrysville clearing, hogs fattened themselves on the mast produced by oak and hickory trees. As new settlers arrived and began transforming more of the forest into field, a traveler on foot could begin to see from the heights of one of the ridges rising up along the Mohekan the graceful silver curves of Fortescue Cuming's imagination.[1] Manchester was undergoing a transformation as well. Once a medieval market town, it was now a cramped industrial city of over 150,000. Thousands of men, women, and children worked twelve- and fourteen-hour days in its steam-powered textile mills. Thick smoke from the soft coal burned by the factories and in the fireplaces of the city's residents hung low in the air and stained the old Tudor storefronts, the churches, the row houses, and the new factories, which lined the narrow crooked streets. Through this miasma flowed the Rivers Irwell and Irk, their waters turned black by coal ash and sewage. "From this filthy sewer pure gold flows," Alexis de Tocqueville wrote a few years later. "In Manchester . . . civilised man is turned back almost into a savage." Another observer noted that "it is scarcely in the power of the factory workman to taste the breath of nature or to look upon its verdure."[2] Both Perrysville and Manchester were undergoing rapid change, but it would be difficult to imagine two more dissimilar scenes in the English-speaking world.

In January 1817, John Chapman was holed up somewhere for the winter, perhaps boarding with Eben Rice and his family along the banks of the Black Fork, anticipating spring and another season of apple tree planting. At the same time the Reverend John Clowes of St. John's Church, Manchester, gathered with a small group of educated middle- and upper-class Mancunians for a regular meeting of group they had formed to promote the religious writings of a deceased Swedish physicist. They called themselves the Manchester Society for the Printing, Publishing, and Circulating of the Writings of Emanuel Swedenborg, a name that may have contained more words than the society contained members. Their task this day was to compile a report of developments that had occurred over the last year. Among the items they included in this annual report was a copy of a letter they had received from a member of a satellite community of Swedenborg enthusiasts in Philadelphia. The letter told a fantastical story:

> There is in the western country [of the United States] a very extraordinary missionary of the New Jerusalem. A man has appeared who seems to be almost independent of corporal wants and sufferings. He goes barefooted, can sleep anywhere, in house or out of house, and live upon the coarsest and most scanty fare. He has actually thawed the ice with his bare feet.
>
> He procures what books he can of the New Church; travels into remote settlements, and lends them wherever he can find readers, and sometimes divides a book into two or three parts for more extensive distribution and usefulness. This

man for years past has been in the employment of bringing into cultivation, in numberless places in the wilderness, small patches (two or three acres) of ground, and then sowing apple seeds and rearing nurseries.

These become valuable as the settlements approximate, and the profits of the whole are intended for the purpose of enabling him to print all the writings of Emanuel Swedenborg, and distribute them through the western settlements of the United States.[3]

The report from Philadelphia did not mention the "extraordinary missionary" by name, this strange primitive man who was busily extending the work of the Manchester society on the edge of Christian settlement in the New World. But the members of the society, who had up to this point found their efforts ignored—or worse, mocked—by most of England's religious establishment, some of whom declared Swedenborg a madman, were certainly heartened to learn of this solitary American's devotion to their common cause. It seemed to them that most men simply did not have the patience to read Swedenborg's writings and penetrate the profound truths they contained. But at least one strange frontier primitive did. And if such a simple man, living on the remote edge of Christian settlement in the New World, could grasp the truths of the Swedish seer, then perhaps the day when all Christians would come to recognize them was not so far off, and a New Jerusalem on earth would be born.

That John Chapman, son of a poor tenant farmer, would become devoted to the spiritual writings of a Swedish nobleman who died two years before his birth is something of a puzzle. The two men had little in common. Emanuel Swedenborg was a man of great intellect and learning; John Chapman had just a common school education during his Longmeadow boyhood. Swedenborg, the son of a bishop in the Swedish state church, was born into a privilege that permitted him to devote his life to study. The writings Swedenborg left behind at the end of his life fill small libraries; John Chapman's copper-plate script only survives on a handful of IOUs and land deeds. Both men remained bachelors for their entire lives, a state perhaps brought on by their individual peculiarities and obsessions. As a man of great talent from a well-to-do family, Swedenborg certainly suffered no material obstacles to matrimony. But his most intimate relationships were with his journals. He devoted his early life to reading and writing on a range of topics including science, mathematics, engineering, and political philosophy. But in his fifties he began having dreams and visions that prompted a profound spiritual crisis. Visited by angels who told him that he had been chosen by God to advance a new understanding of the scriptures, Swedenborg abandoned his scientific studies and devoted his days to spiritual writings and scriptural exegesis. The angels also allowed him to travel to heaven and hell and revealed all the secrets of the afterlife to him. Swedenborg's descriptions of heaven and hell were vivid and detailed, but he insisted that he alone had been granted this ability to converse with angels and visit the spirit world. He sternly warned others not to attempt such contact.[4]

Swedenborg's stories of his conversations with angels and his visits to heaven were not the first John Chapman had encountered. He surely heard the stories of the prophet Handsome Lake's travels to the spirit world when he was living just a few miles away from his village of Burnt House in northwestern Pennsylvania. Those stories, in fact, had many similarities to Swedenborg's. And the spiritual visions of Tenskwatawa and the story of how he predicted a solar eclipse were well known in the settlements along Owl Creek and the forks of the Mohican. Furthermore, the trans-Appalachian frontier in the first decades of the nineteenth century was crawling with would-be prophets. Some, like the Leatherwood God of Guernsey County, Ohio, claimed to be God in person. That John Chapman became drawn into Swedenborg's heavenly visions made him neither particularly gullible nor insane, as some have suggested. He was in good company.

Swedenborg's visions and conversations with angels led him toward a theology that upended many of the assumptions of the major Christian churches of his day. Among these were that the Second Coming had occurred in 1757, not in the physical world but in the spiritual one; that one's spiritual fate was not sealed upon one's death, but moral progress or decline could continue in the afterworld; that the dead could choose to dwell in heaven or hell, and some troubled persons might prefer the latter; and that others might be content to dwell in the lower levels of heaven forever, but some would seek to advance to higher places. Swedenborg also rejected an idea fundamental to the Lutheran Church—the state church of Sweden—that salvation was achieved by faith alone. Real faith was manifested in acts of charity and commitment to a life of usefulness.

While both the Catholic Church and the dominant Protestant ones promoted a God-centered heaven where physical needs and desires did not exist, and therefore marriage did not exist, Swedenborg presented a human-centered vision of heaven where the marriage bond continued. What Swedenborg called "conjugial love"—a perfect love between two persons who were true soul mates—existed in heaven, but people's heavenly partners might not be the same as their earthly ones, especially if their earthly marriages had been less than ideal. Those who were celibate on earth, and for some reason chose to maintain that state in heaven, were segregated to a part of heaven where they might retain that "unnatural" state. But lifelong bachelors like himself, who desired an ideal "conjugial" relationship in the afterlife, would be partnered with a true soul mate in heaven.[5]

Finally, Swedenborg's "Doctrine of Correspondences" asserted that everything in the physical world had a spiritual analog, an insight that led him to declare that even the simplest, most direct passages of the Bible contained hidden meanings that had become clouded and lost by Old Church leaders. He promised that his methodical biblical exegesis recovered these meanings, and once members of the Old Church had read and embraced them, a New Church—a Church of the New Jerusalem—would emerge on earth, bringing the Second Coming that had already occurred in the spirit world out into the physical one.[6]

Swedenborg devoted the last twenty-eight years of his life to producing sixteen books outlining his visions of the afterlife, as well as several volumes of scriptural interpretation. He published his writings at his own expense. They were mostly ignored and sometimes ridiculed, but eventually the Swedish state church pursued heresy charges against some of his champions. Swedenborg never despaired. He knew that it might take some time for their truths to be absorbed and accepted. He seemed content to simply continue producing them, until his last days. Labeled a heretic in his home country, he spent his last years in Amsterdam and London, where he died in 1772.

Even in the early years after his death, Emanuel Swedenborg's writings attracted only scant attention. In Manchester, the Reverend John Clowes was among the first to be persuaded by Swedenborg's radical reinterpretations of scripture. Clowes began translating them into English and formed one of the earliest Swedenborgian reading groups in 1778.[7] The tiny Manchester society was joined by an equally small group of Londoners a few years later. Some, like Clowes, were content to continue spreading Swedenborg's ideas while remaining within the Church of England. Swedenborg never called for the formation of a distinct sect, expecting his ideas to gradually penetrate the minds of the leaders of the established ones. But others felt constrained operating within the bounds of the established church and formed the Church of the New Jerusalem in the hopes it would advance Swedenborg's ideas more rapidly.[8]

PHILADELPHIA

The New Church, as it came to be called, attracted enthusiastic converts only gradually. In 1781, Scotsman James Glen, owner of a plantation in British Guiana, became a convert after reading Swedenborg's *Heaven and Hell* in Latin while on a ship at sea. Three years later he had sold his South American plantation, relocated to Philadelphia, and placed a small ad in a local paper announcing the formation of a group to read the works of Swedenborg.[9] Among the earliest participants in Glen's small group was Francis Bailey, a Philadelphia printer, who in 1787 printed and distributed freely John Clowes's "A Summary View of the Heavenly Doctrines." That tract drew enough interest for Bailey to launch a plan to publish all of Swedenborg's works in English, and by 1789 he had fifty subscribers to the venture, including Benjamin Franklin. It was a promising start but not enough to support the costs of publication, and Bailey devoted much of his personal fortune to getting just a few English translations of Swedenborg into print.[10]

Bailey was able to recruit other men of capital to the cause. Philadelphia merchant William Schlatter was so enthusiastic about the possibilities of print evangelism that he spent most of his personal fortune on the effort. By 1817, he had distributed over three thousand books and tracts by inserting them gratis into orders of cotton and calico sent to country merchants. By 1819, that number had reached seven thousand. While the New Church was not seeing the dramatic conversion numbers that the Methodists and other sects who appealed to emotion and the spectacle of the revival were achieving, devoted New Churchmen continued to be optimistic. Schlatter believed that in the trans-Appalachian West, the New Church would eventually yield "a great harvest . . . we may reasonably expect, whenever they do get into an enquiring state, that our doctrines will suit them better than the Old Church, for they are an independent, free-minded people, and not disposed to be shackled in religion or politics."[11] Awakening to Swedenborgian truths would not occur like a bolt of lightning from the sky but as a gradual enlightenment brought on by careful and extensive study.

While evangelism through the printed word may not have been the most rapid way to win converts, it was probably the most effective way to disseminate the complex ideas of Swedenborg. It is also likely to have been the tactic that won John Chapman as a convert. Earlier scholars of have speculated on the question, "Who introduced Swedenborgian ideas to John Chapman?" Some have concluded that it was probably Judge John Young of Greensburg, in western Pennsylvania, as he was one of the earliest converts to the New Church west of the Alleghenies. But these scholars might have been asking the wrong question.[12] Many of the most enthusiastic promoters of the ideas of Swedenborg came to embrace his ideas through the solitary experience of reading and contemplation. Some, to be sure, were introduced through social connections or participating in a reading circle in the parlor of a middle-class home. But even in those cases, true conviction emerged gradually through careful reading. It seems unlikely that the ragged John Chapman first encountered Swedenborg's ideas in a parlor reading circle. He was by no means antisocial, but his appearance placed him outside the educated middle-class circles from which most New Churchmen came. Furthermore, he spent a great deal of time alone and may have had a passion for reading, as his purchase of "two small histories" from John Daniels's dry goods store in 1797—the only sale of books, save a few Bibles, recorded in the first five years of settlement along the Brokenstraw—suggests. It is not hard to imagine that his first encounter with Swedenborg came in the form of one of William Schlatter's books, which arrived on the frontier tucked into a bolt of calico shipped to a country store and was sold or given to Chapman by an indifferent country merchant. If James Glen's conversion occurred by dim candlelight in the dark hold of a ship traversing the Atlantic, perhaps John Chapman's enlightenment came while the barefoot wanderer rested against the trunk of a tree in the wilderness. Schlatter no doubt included contact information for the Philadelphia New Church in the books and tracts he distributed across the West, and he was the earliest contact between John and the New Church that can be documented. By the spring of 1817, Schlatter had sent some Swedenborgian works directly to John Chapman, but his familiarity with Chapman at that time was still novel.[13]

While John's first exposure to the ideas of Emanuel Swedenborg was probably through print, he eventually made contact with New Church missionaries. In 1817, in an effort to emulate their more successful rivals the Methodists, the Philadelphia New Church Society sent out two missionaries on a thirty-nine day tour of the South and West during which they baptized thirty-seven converts. The two men apparently encountered John Chapman on this trip and described him as an intelligent and zealous champion of Swedenborgianism. The following year, New Churchmen David Powell and Thomas Newport did an eighteen-stop tour of Ohio, preaching to "large audiences." John Chapman, who attended outdoor preaching events whenever he could, was likely among the crowd at one or more of these stops. In 1819, Thomas Newport hosted an open-air revival style meeting on his own farm in Lebanon, Ohio, north of Cincinnati, to which two to three hundred people attended. "More interest I never saw, nor better behaviour... the sphere of love was ecstatic... the people are like a ripe harvest," Newport optimistically reported. Nonetheless, the meeting yielded just nine new communicants.[14]

Westerners were open to listening to the ideas of Emanuel Swedenborg, but they were not proving to be easy converts. The emotional style of the Methodists and Baptists and the egalitarian spirit of their often poor, simple preachers made inroads into the ranks of Ohio's unchurched very quickly. Abraham Lincoln captured the attitude of many westerners a few decades later when he remarked "When I hear a man preach, I like to see him act as if he were fighting bees."[15] Swedenborg's complex doctrines could not be easily adapted to the Methodist style. And his vision of the afterlife, in which moral progress could occur after death, meant that frightening listeners with hellfire and damnation, such an effective method for the circuit-riding preachers, was not a strategy New Church preachers were likely to adopt. Swedenborg offered up a religion based more on hope than on fear.

New Churchmen heeded Emanuel Swedenborg's warning that it would take time for the deep truths in his writings to penetrate the minds of most of Christendom, but there were some reasons for them to be optimistic about their prospects in the United States and in particular in the lands west of the Alleghenies. While new ideas about God, heaven, and the path to salvation were emerging on both sides of the Atlantic, in the United States, and in particular the new settlements of the interior, Swedenborgians and other religious innovators faced less resistance. In England, the French Revolution heightened fears of radicalism and brought on a new conservatism in the state. The established church and the state itself cast a wary eye on new religious ideas and actively sought to control and limit them. Anyone preaching outside of the established church still had to apply for a dissenter's license.

In the new United States, by contrast, where established churches at the state level were mostly weak or nonexistent, even the word "dissent" had ceased to have much meaning. Many religious ideas with their roots in England or Europe found new adherents in the young United States or were even remade by American adopters.[16] At one end of English dissent stood the English Methodists, struggling to revitalize the Church of England from within but moving cautiously to demonstrate their loyalty to the official church and the state and often acting as the most vocal critic of more radical kinds of dissent. It was the English Methodists, after all, who first took on the radical doctrines of Swedenborg and tried to discredit him by claiming to have evidence that he was insane. Methodists had some early success in the British North American colonies, but after the Revolution, American Methodism, freed from the need to demonstrate allegiance to the government and the established church, exploded into the most successful religious movement of the age. At the other end stood groups like the United Society of Believers in Christ's Second Appearing, commonly called the Shakers. Founded in England in the middle of the eighteenth century, the Shakers presented a much more dramatic challenge to the English social order. They disposed of their worldly possessions and embraced communalism, celibacy, and social and gender equality. In 1774, Shaker leader Mother Ann Lee, "the mother of the new creation," migrated to the United States where the Shakers found more success than they had in England, planting more than twenty colonies and attracting more than twenty thousand converts in the nineteenth century. The success of the Methodists and Shakers in America gave the devotees of the writings of Emanuel Swedenborg, and their most famous convert, John Chapman, reason for optimism about the reception of new truths in a new land.

The vast numbers of unchurched on the frontier presented an opportunity for the new evangelizing sects, but not necessarily an easy one. The two most common religious identities west of the Appalachians were, according to one traveling minister, "nothingarian" and "anythingarian."[17] Freed from the conservative restraints of an established church working hand in hand with the state to enforce order and conformity, Ohio's new emigrants were open to considering all kinds of new religious ideas, or none at all, without stigma. And they also felt free to change their minds. Colonies of the United Society of Believers became accustomed to taking in "Winter Shakers"—desperate people who declared themselves true believers as the weather turned cold and enjoyed the food and shelter provided by the community during the shortest days of the year but departed from the stifling discipline and self-denial of Shaker communities as soon as spring came and life on their own was easy again. Even the Methodists, whose numbers were increasing by leaps and bounds, understood that many of those who experienced the emotional conversion experience of the camp meeting would soon return to their reprobate ways. New Churchmen, too, found some enthusiasts departing for other isms, even as new ones entered their reading circles. And in this open atmosphere, in which there were few restraints on individual expression, splinters and schisms became a constant reality for almost all religious groups in America. Americans simply no longer felt constrained to accept one view or interpretation uncritically, and their minds were quite open to impassioned persuasion. Furthermore, they seemed to reserve the right to pick and choose ideas about heaven, hell, and salvation from an à la carte menu.

What other religious ideas John Chapman might have been drawn to before becoming a devotee of Swedenborg, we cannot know. But once he became a convert to New Church ideas, he did not stray, and for the rest of his life he eagerly shared the ideas of the Swedish mystic with as many willing listeners as he could find. Having acquired a few more books from William Schlatter, he soon emerged as central Ohio's most enthusiastic New Church evangelizer. Many family traditions recall John's appearance at the cabin door, asking the residents if they were ready to hear "Good news fresh from heaven!" When he found a receptive audience, he read aloud to the family around the hearth. At other times he left reading materials behind. But as he was able to carry only a small selection with him, he came into the habit of unthreading the bindings of the books, leaving behind just one section with each family, and promising to exchange it for another section upon his return. Some of these may have been bound volumes of tracts, threaded together in such a way to make them easy to divide and share. If he indeed was splitting apart sections of Swedenborg's longer works, such a strategy was unlikely to win many converts, as understanding New Church doctrines was challenging enough when read in the intended order.[18]

After his first contact with the Philadelphia New Church missionaries in 1817, John maintained an irregular correspondence with the Philadelphia group for five or six years. He was not simply content with proselytizing the individuals and families who took him in but appeared intent to build a real New Church community on the Ohio frontier. By 1820, he had joined forces with Silas Ensign, a former Methodist circuit rider living in Mansfield who had become converted to the doctrines of Swedenborg, and they had formed a small group of Swedenborg enthusiasts. John reached out to his primary contact in Philadelphia William Schlatter, to see if the Philadelphia society could license Ensign as a lay leader.

About the same time, Chapman made William Schlatter an unusual offer. The economic situation in the nation had taken a sharp downturn. A postwar boom fueled in large part by the proliferation of paper notes of questionable value turned to a bust when the Bank of the United States began calling in specie payments. In central Ohio, this meant the rapid evaporation of money. Settlers defaulted on land payments and returned to a barter economy to get by. John was not immune to this crisis and found it difficult to find the cash to buy more Swedenborgian books. So he offered Schlatter a land-for-books swap. He would exchange a quarter section of land he had acquired in Wooster County for as many books as the Philadelphia New Church could provide. But Schlatter, who had emerged as the primary financial supporter of the Philadelphia New Church, was in no position to accept the offer. Western dry goods merchants made up a substantial portion of his clientele, and orders from the cash-starved

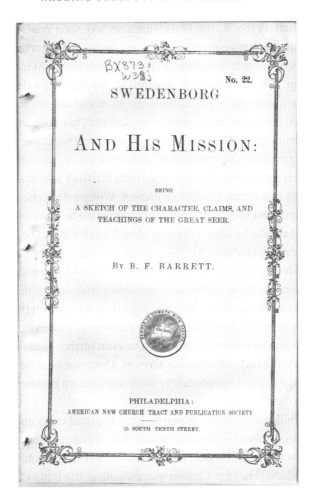

John Chapman spent much of the money he earned from his trees on Swedenborgian tracts like this one. He was known to unthread parts from their bindings and distribute them to frontier families. Swedenborg Memorial Library, Urbana University, Urbana, Ohio.

West were drying up. He faced a cash-flow problem as well. Schlatter advised Chapman to approach the members of the growing New Church Society in Cincinnati, who might be in a better position to take advantage of a land-for-books swap.[19] Within a few years, Schlatter was bankrupt and the society was forced to sell off the temple he had built for it.[20]

To make matters worse, the Philadelphia society soon confronted a rival group of Swedenborgians competing for souls in the city. Mancunian William Metcalfe was a follower of one of the original members of the Manchester Swedenborg group, William Cowherd. Inspired by Swedenborg's description of meat-eating as a dramatic sign of Man's fall, Cowherd and his followers embraced vegetarianism and abstinence from alcohol. They split with the Manchester Swedenborgian society over the issue of vegetarianism and formed their own church under the name Bible Christians. When Cowherd died at the relatively young age of fifty, his critics pointed to his death as evidence that vegetarianism was unnatural. Cowherd's protégé William Metcalfe took up the cause, and convinced forty of his followers to migrate to Philadelphia in 1817, where they established a Bible Christian church in that city. The Bible Christians did not exactly take the city by storm. In fact, Metcalfe lost about half of his followers shortly after arriving. Over time the Bible Christians drifted away from their Swedenborgian origins, and abstinence from animal flesh and alcohol became their central tenets. Metcalfe published sermons on these subjects and gained more notice as the temperance movement gathered steam in the late 1820s. On vegetarianism, he proved to be even further ahead of American reformers. Metcalfe had published his sermon, *Abstinence from the Flesh of Animals* in 1821, more than a decade before Sylvester Graham began championing the vegetarian diet.[21]

With the Philadelphia New Church facing a financial crisis and challenges from other groups like the Bible Christians, John Chapman soon found himself turning elsewhere for support for his mission. In the first half of the 1820s, just as the Philadelphia New Church was dwindling, the Bible Christians experienced modest growth. Faced with falling fortunes, the Philadelphia New Church was able to offer John Chapman little help in his effort to spread the word in central Ohio.

CINCINNATI

Schlatter had specifically recommended that Chapman contact Wright and Marcus Smith, two of ten brothers, each about six feet tall, who had recently moved to Cincinnati and come to be known as "the Sixty foot Smiths." At least five of the brothers were enthusiastic New Churchmen, busy promoting Swedenborg's doctrines in the rapidly growing Ohio River city. Manchester native Adam Hurdus arrived in Cincinnati in 1808 and by 1811 had formed a society with seventeen or eighteen members. In 1819, the Hurdus group built a New

Church temple that could seat 350 persons. By 1822 the Cincinnati New Church was the center of communications for other Swedenborgians in the West. That same year the group took a turn toward anticlericalism, an increasingly common phenomenon in American churches in this democratic age. Infected perhaps by the egalitarian spirit of the Methodist and Baptist movements, some Cincinnati Swedenborgians began to question whether ordination and licensing from a higher body should be a qualification for preaching. Of the many enthusiastic Swedenborgians who preached in Cincinnati in the 1820s, only Adam Hurdus was ordained, and even he appears to have come to doubt its necessity.

Cincinnati was also emerging as a center of reformist impulses, and the city's New Churchmen were at the forefront of some of the more radical reform experiments. Scotsman Alexander Kinmont, who had married Francis Bailey's daughter arrived in Cincinnati with ambitious plans to establish a new school based on his unique theories of education. John Chapman, who had already formed a relationship with the Baileys, became a friend of Kinmont as well. But in 1824, enthusiasm for radical reform threatened to disrupt Cincinnati's New Church altogether after a number of Swedenborgians became enamored of socialism after hearing Scottish socialist utopian Robert Owen speak in the city. They decided to establish their own new society in the Ohio wilderness. Owen's own utopian visions were secular, but he had much hope for the Swedenborgian socialist utopia he had inspired, believing their common faith and intellectual discipline would make them well suited for a utopian venture.[22]

The group secured a tract of land at Yellow Springs, seventy-five miles north of Cincinnati and persuaded about one hundred colonists to resettle there. Believing that they would soon be flooded with additional applicants, they agreed to cap the colony's population at two thousand. Things began to sour very quickly. Some in the group soon decided that the original plan of perfect equality in housing for all was impractical and argued that some should have better housing than others. Laborers in manufacturing argued that the value of their products exceeded those of the agricultural laborers so they should only have to work for half a day while farmers worked a full one. By Christmas of the community's first year, most of the Yellow Spring colonists, concerned about spending a harsh winter in primitive shelters, had packed up and returned to the comforts of Cincinnati. The colony was completely dissolved within a year, although at least one of its members, Daniel Roe, who had been a leader in Cincinnati's New Church community, went west to join Robert Owen's New Harmony community on the Wabash.[23] If John Chapman visited the Yellow Springs colonists, no record remains of it. But he had by 1824 established connections with some of the people who launched it. Utopian experiments challenging the economic and social status quo were a part of the American landscape in these years. Groups that sought to reorder the relationships between capital and labor, between men and women, and to alter the customs of economic exchange were constantly springing up. John Chapman's personal experiments in living outside the rules of common society must be understood in this context.

The efforts of Cincinnati's New Churchmen to promote Swedenborg's ideas were only temporarily derailed by the Yellow Springs experiment. They embraced every opportunity to share Swedenborg's wisdom. A Universalist newspaper in the city agreed to publish a few sermons by New Churchmen, prompting a spirited debate. One skeptical respondent declared Swedenborg to be "insane" and offered a patronizing, but insightful, description of New Churchmen. They were "a liberal, good-natured, cheerful sort of folks, not very profound indeed, but on whom their religion, such as it is, made no other bad impressions than excessive vanity, and obstinancy . . . But they have none of the cant, none of the moroseness, none of the affected airs of superiority which distinguish the real true blue orthodox. They can all enjoy the good things of life, without murmuring, and muttering, and making long faces; and if they have a dash of the ludicrous now and then, in divine human metaphysic-long lucubrations, ascending in giddy heights through many heavens it is not to be wondered at, considering the daring heights to which they presume, as earthly guests, thinking themselves inspired, with true 'empyreal air.'"[24]

The condescending description might have a ring of truth when applied to the middle-class participants of Philadelphia, New York, Boston, and Cincinnati Swedenborgian reading circles. But in this age of expanding individualism, even *choosing* the tiny New Church was an act that suggested a predilection toward independent-mindedness. John Chapman may not have been a "typical" Swedenborgian, but the group did seem to attract people ready to break from the social expectations of the wider society. Robert Carter, of the wealthy Virginia Carter dynasty, had embraced Swedenborg's theology by this time and was the first Virginia slaveholder to take the radical action of freeing all of his slaves during his lifetime, an act that utterly impoverished him. He spent his last days as a tenant in a Baltimore rooming house. Swedenborg's writings had also inspired the vegetarianism of William Metcalfe. If many New Churchmen only had brief flirtations with radical reinvention, as the Yellow Springs group had, the New Church still attracted quite a few people willing to deprive themselves of great comfort for their principles.

In the memories of those who encountered him, John Chapman was frequently described as a "rigid Swedenborgian." In debates on the nature of heaven, and on human salvation, Chapman seemed consistently to take the Swedenborgian view, which he understood deeply. Given his simple education and his primitive lifestyle, it would be easy to assume that Chapman became drawn in by Swedenborg's vivid descriptions of the afterlife and had only a limited grasp of his theology. But the evidence suggests John's attachment to New Church theology went deeper than just that. He understood even the finer theological distinctions that separated Methodists, Quakers, Universalists, New Churchmen, and others.

One story recounts an overnight visit Chapman had with the Quaker Roberts family of Knox County in 1819. John, ever eager to read, noticed that the family had a book by Hosea Ballou, *On Atonement,* and after supper asked if he could read it. Ballou was the leading propagator of universalism in America in these years—the idea that all souls would eventually be saved, a doctrine anathema to Methodists, Baptists, and Presbyterians. Chapman opened the book "and read for some time by candle-light, thinking it at first it was good Swedenborg, and desired to take it with him; but after he read further, and found the kind of doctrine that it inculcated, he threw it down indignantly, expressing his disappointment, and in a few moments after stretched himself out and went to sleep."[25] Many educated people today could not, after reading Ballou and Swedenborg, articulate their differences on the issue of salvation, and some have lumped the Swedenborgians into the Universalist camp. That John was able to recognize both the similarities and the differences in the doctrines after reading a few pages, suggests that he was well steeped in New Church doctrines. When it came to discussions of theology, John Chapman was no lightweight. The ministers and other deeply religious people he encountered had great respect for his intellect, even when they rejected his beliefs.[26]

Those who were indifferent or not deeply committed to religious ideas sometimes found John's enthusiasm wearisome. One surviving story from Mansfield, Ohio, recalls that in 1825, a local judge named Wilson, tired of John's lengthy ramblings on Swedenborg, suggested that an Irish servant girl, the cook at a local tavern, was a promising object for conversion. John took up the idea and spent days pestering the poor girl, explaining Swedenborg's doctrines while she busied herself churning butter and engaging in other daily chores. The young woman eventually lost her patience and grabbed a fire stick, waving it at Chapman and ordering him to "Clare ute from me with yur religions, you good for nothing spalpeen."[27]

Others who only half-listened to his explanations of heaven and hell dismissed them as the ranting of a lunatic. And it is no surprise that many who encountered Chapman attributed, often incorrectly, his behaviors to his Swedenborgian religious convictions. Among those was his lifelong celibacy. It was not unusual for religious evangelizers in the early nineteenth century to remain bachelors. In fact bachelorhood had been the common state of most of the early Methodist circuit-riding preachers. The pay was so low and the conditions so rough that the church discouraged marriage. Many of the circuit riders eventually did marry, but it usually meant the end of their circuit-riding days. In general, however, as Ohio became domesticated, people tended to view bachelorhood as an unnatural state, and one increasingly associated with immorality. Bachelors began feeling the need to explain their unnatural condition, or at least those around them did. When bachelor James Buchanan ran for president in the 1850s, his political allies thought it necessary to spread a story about how he lost his one true love as a young man, that his heart never healed and he could not give it to another. Such stories were meant to invoke sympathy and admiration and to derail any suggestions that the bachelor status was the result of a poor moral constitution. To the contrary, the heartbroken bachelor who committed to a solitary life was the ultimate romantic soul, one whose commitment to the perfect union between man and woman left him with no purer choice when thwarted in his attempt to find true love. Stories like these became a part of the Johnny Appleseed legend. Perrysville native Rosella Rice, who encountered John Chapman when she was a little girl and he a guest in her family home, wrote years later that "we had always heard that Johnny had loved once upon a time, and that his lady love had proven false to him." Other stories told of a love back in Massachusetts who had been forced to marry someone else. And when historian Henry Howe put the Johnny Appleseed story in the 1847 edition of his *Historical Collections of Ohio*, he reported that Chapman proposed to a "Miss Nancy Tannehill," of Perrysville, but was too late, as she was already engaged.[28]

Others saw his celibacy as another manifestation of his life of self-denial and tried to connect it to his Swedenborgian beliefs. One tall tale that made the rounds claimed that John had been visited by two female angels who promised him they would both be his wives in heaven if he remained chaste on this earth. When two boys asked Chapman if he planned to sleep with these wives in heaven, his alleged reply was, "Oh Yes! My two wives will sleep in a bed, and I intend to have a wide board planed smooth and nice attached to the side of the bed for me to sleep on!" The joke pokes fun at two of Chapman's notable peculiarities, his celibacy and his preference for sleeping on hard floors. The tale is of course a gross misrepresentation of Swedenborgian doctrine, but it contains just enough elements of New Church ideas—that Swedenborg was visited by angels, that marriage would exist in heaven—to suggest that the original spinner of the tale had one ear open when John was explaining his religious views. Other versions of the tale said that John went without a wife on this earth so that he might have a "pure wife" in heaven.

Chapman's lifelong celibacy was not derived from any instruction he had received in Swedenborg's writings, but it was perhaps tied to his attraction to Swedenborgianism in more complex ways. Both Swedenborg and Chapman were lifelong bachelors, but in each case the failure to find a spouse on earth was not due to a belief that celibacy in this world would earn them special rewards in the next one. New Church teachings idealized the marriage bond to a greater degree than most other Christian sects. Swedenborg's highly idealized visions of a perfect marriage in heaven are not surprising when we know of his own infatuation with a married woman, the Countess Gyllenborg. In one of Swedenborg's visions it is revealed to him that the countess is his true soul mate–the person he will spend eternity with—and her earthly husband, Baron Frederick Gyllenborg, appears in heaven briefly in the body of a cat—the form often taken by the morally inferior—before he decides to dwell in hell forever.[29]

That very idealization of the perfect "conjugial" relationship between soul mates may have been one of the central aspects of New Church doctrine that first appealed to John and also made his search for a partner more difficult. He was too young when his mother died to have any direct memories of her. Instead, she existed only in family stories and in her final letter to John's father. In that letter passed down through the family for generations, Elizabeth Simonds Chapman existed as the most pure and saintly marriage companion. As she lay on her deathbed, she wrote of her desire to "glorify God here and to finally come to the enjoyment of Him in a world of glory" and of her final hope that "we shall both be as happy to spend an eternity of happiness together in the coming world which is my desire and prayer."[30] No flesh and blood woman that John Chapman would ever encounter could possibly meet the standard of pure love and perfect piety embodied in that letter.

Add to this Chapman's many peculiar character traits, and it is no surprise that true love evaded him. His rough and bedraggled appearance and many of his peculiar behaviors might be reformed by the right partner, but other aspects of his personality set him apart from the typical frontier male. Even those who knew him in Pennsylvania commented on his extraordinary meekness, and that many women on the Ohio frontier did not really view him as a conventional man. One story recounted by Rosella Rice is especially revealing on this point. One day when she was a young girl, and her mother was ill and not dressed for company, Rosella and her siblings spotted a man coming toward the house. They warned their mother of the impending arrival of the strange man, "but when [mother] saw him through the window she said, 'Why it's dear old Johnny' in a voice that showed how glad she was to have him come when she was ill—satisfied as if he had been a good woman coming down the path." The John Chapman that Rosella Rice remembered "liked women better than men. He seemed feminine in many of his attributes, and in his likes and dislikes he was decidedly womanish."[31] When Chapman arrived at a pioneer home, Rosella recalled, he seemed instantly attracted to the company of women and eager to learn about the development of each child since his last appearance. All of these factors combined—his peculiar habits, his "womanish" traits, the frozen image of his mother as the perfect woman, and the idealization of the marriage bond encouraged in New Church writings—may have made the odds of John finding a partner in this world too long.

But some stories suggest he never gave up on the idea of marriage here on earth. His strategy for finding a pure wife was, if the many stories are to be believed, an unconventional one. The "Miss Nancy Tannehill" of Perrysville mentioned in Howe's history was born in 1807. She would have been somewhere between eight and twelve, and John about forty, when he sought her hand. A Tannehill family story recounts that John in fact asked her father Melzar Tannehill if he could marry his young daughter once she was older, but Melzar lied and told him "she was already taken." If the family story is to be believed, Melzar was more concerned with the fact that John was a "pedlar" than the age difference. Years later, when Nancy Tannehill was in her twenties, she married a doctor instead. But this is not the only story about John's efforts to secure a marriage contract with a young child. One claims that he doted on one young girl and provided for her for years, in anticipation that this child would one day be his perfect wife. When, on one visit, he witnessed the girl in flirtatious conversation with a young man, he left in anger, and never spoke to her again. In another version of the story, John gave up on the girl when he witnessed her playfully kiss a neighbor boy on the cheek. As late as 1830, one person claimed to see a fifty-something-year-old John traveling in Crawford County, Ohio, in the company of a little girl he called his "angel."[32] Such stories repeated today bring the modern hearer great discomfort, but they were told in the mid-nineteenth century with no suggestion of immorality. John Chapman was recalled affectionately by many central Ohio families, and there is no indication that any ever considered him a threat to their children.

THE PRIMITIVE CHRISTIAN

John Chapman's exceptional generosity and his commitment to leading a life of usefulness would certainly find support in the words of Emanuel Swedenborg. Swedenborg's insistence, for example, that faith was by necessity manifest in acts of charity encouraged his followers to pursue lives of benevolence. But there were many aspects of John Chapman's life and behavior that could not find justification in the ideas of Swedenborg or the New Church, although many who encountered him assumed they did. Among these were his radical commitment to living the life of the primitive Christian. John Dawson, who knew Chapman in his last years, claimed that he "devoutly believed that his physical mortifications on earth insured him a greater fullness of celestial bliss; hence his self-denial as to personal comforts and wants."[33] Other early chroniclers claimed that Chapman believed that his self-imposed austerity would earn him "snug quarters hereafter."[34] But such ideas are hard to find in the writings of Emanuel Swedenborg. Swedenborg did not deny himself material comforts, and he rejected the idea that a good Christian needed to live an ascetic life. "The rich enter heaven as easily as the poor," he insisted, and in one of his descriptions of the afterlife, he noted that people who deprived themselves of comfort on earth because they believed it would gain them special status in heaven "are of mournful character" in heaven. "They are resentful when they do not receive happiness beyond the lot of others, believing they have earned it."[35] It is worth treating the oral tradition's frequent association between Chapman's self-denial and his expectations of heavenly rewards with a dose of skepticism. It is not surprising that those who chose to live with more material comforts, in struggling to understand Chapman's seemingly irrational behavior, might impose such an explanation on it. Certainly thrift and frugality were virtues embedded in the Protestant tradition. But the words "thrift" and "frugality" are inadequate for describing Chapman's extreme commitment to self-denial. John certainly found the New Testament an inspiration, living quite literally the words of Matthew 6:25, "Therefore I say unto you, take no thought for your life, what ye shall eat or what ye shall drink; nor yet for your body, what ye shall put on. Is not the life more than meat, and the body more than raiment?" He appears to have taken this verse to heart, to an extent that few Christians have done.

This commitment to a life of extreme austerity and self-denial was the source of most of the tall tales that emerged around him. It is impossible for a historian dependent largely on oral traditions to draw a firm line between truth and folklore when it comes to stories about Chapman's primitive dress, diet, and determination to forgo even basic comforts. Chapman's preference to go barefoot whenever possible, for example, is an element of almost every oral tradition about him, yet the stories range from the probable—that he went without shoes during the warmer months of the year—to the fantastic—that he walked across a frozen Lake Erie barefoot or thawed ice with his bare feet.[36] Sorting truth from fiction in this case is an exercise in setting the historical context and dealing in probabilities, not certainties. Barefootedness under certain circumstances was a common condition on the Ohio frontier. Shoes were a rare and valued item in the early days, so many settlers spared their shoes when they could, in order to extend their life. That the first story of John Chapman's ability to "thaw the ice with his bare feet" made it into a New Church report written sometime in 1816, however, suggests that Chapman's commitment to the barefoot life was clearly much more profound than that of his frontier neighbors, or else it would not have become fodder for a tall tale.[37] Interestingly, the stories of Chapman's unshod feet are a part of the Pennsylvania, Ohio, and Indiana traditions and continued to be recounted throughout the rest of his life. Judge Lansing Wetmore of Warren, Pennsylvania, had claimed that Chapman crossed the Alleghenies barefoot when he was just twenty-two or twenty-three, and the writer of Chapman's obituary also noted that "in the most inclement weather he might be seen barefooted and almost naked."[38]

Despite the number and universality of these stories, there is evidence that Chapman sometimes wore shoes or foot coverings of some sort. He purchased several pairs of moccasins at the Holland Land Company store in the 1790s.[39] Others commented that he sometimes wore sandals in the summer, when rough roads hurt his feet. Another claimed he never purchased shoes but did make use of shoes discarded by others. "When he used anything in the form of boots or shoes, they were cast off things, or generally unmated, which he would gather up, however dilapidated they might appear—always insisting that it was a sin to throw aside a boot or shoe until it had become so thoroughly worn out as to be unable to adhere to a human foot."[40] Elias Slocum, who encountered Chapman wearing shoes that were falling apart one late November, insisted he take a spare pair he had that did not fit him. A few days later Mr. Slocum encountered a barefoot Chapman walking down the snow-covered streets of Mansfield, and when he demanded an explanation, "Chapman replied that he had found a poor, bare-footed family moving westward, who were in much greater need of clothing than himself, and that he had made a permanent present of them."[41]

Chapman also apparently had no fear of walking through the forests without protection for his feet. When Josiah Thomas asked him if he was afraid of being bitten by poisonous snakes as he trod through the woods, Chapman allegedly held up a copy of a New Church book, or perhaps the New Testament, and responded, "This book . . . is an infallible protection against all danger, here and hereafter."[42] One time when he was struck by a serpent, he lashed out in the moment of pain, and drove his hoe through the snake. He immediately felt remorse for allowing his emotions to rule him and deeply regretted the death of the innocent reptile.[43] Years of harsh treatment had made his feet "dark, hard, and horny," recalled one who knew him.[44] In general, his response to pain was that of a Stoic. Rosella Rice, who knew him when she was a little girl, claimed that he "gloried in suffering" and that "if he had a cut or sore, the first thing he did was to sear it with a red hot iron, and then treat it as a burn."[45] He also was known "to thrust pins and needles into his flesh without a tremor" to demonstrate to admiring children, perhaps, his ability to endure pain.[46]

While he was frequently a guest at the homes of others, when time came for sleep, he was notorious for declining the offer of a bed, preferring to sleep outdoors when the weather permitted, or on the hard puncheon floor. He may have occasionally relented and accepted the comfort of a straw mattress as he aged, for one Indiana resident who recalled Chapman's visits to his father's house when he was a boy said, "[H]ow I hated to sleep with that old man! I never wanted to, but Dad made me."[47] Yet other acquaintances who knew him in his last years don't ever recall him sleeping in a bed; instead, "he preferred to lie on the floor of a tavern or private house—always laid in the bar-room of a hotel, when stopping there, and when necessary, kept fire during the night."[48] Yet in his estate papers several claims for unpaid room and board were made.[49]

In his clothing, too, John Chapman began to draw attention as the country became more settled and middle class. The famous story of a garment made from a coffee sack, with a hole cut for his head and two for his arms was recorded in some of the earliest print stories.[50] Even those who remembered him from his Pennsylvania days recalled his clothing as being well worn. "I do not know that anyone ever stole his coat; the truth is he seldom ever had one worth stealing," R. I. Curtis recalled.[51] Others commented that even in cold weather, he rarely had a coat. Salathiel Coffinbury, who knew Chapman in Mansfield, said he never saw him in a coffee sack, but he generally wore "the off cast clothing of others." He was "often in rags and tatters, and at best in the most plain and simple wardrobe [but] he was always clean, and in his most desolate rags comfortable, and never repulsive." Coffinbury also claimed that his mother once constructed for John a coarse tow-linen coat according to his instructions. "This coat was a device of his own ingenuity and in itself was a curiosity. It consisted of one width of coarse fabric, which descended from his neck to his heels. It was without collar. In this robe were cut two arm holes into which were placed two straight sleeves."[52]

Chapman was also regularly innovating in head gear. The story of the tin pot for a hat was told early and often; though how often he wore this on his head is questionable, as anyone willing to experiment with such headgear will attest. For a few years after the war, one person claimed he wore a fancy military chapeau given to him by a returning officer. Others encountered him wearing a homemade cap with a long pasteboard bill to protect his eyes from the sun. It appears that his headgear was frequently changing but usually noteworthy.[53]

In diet as well, Chapman was austere and peculiar. A Pennsylvania legend suggested he subsisted for an entire winter on nothing but butternuts. A day laborer John once hired to help him clear some land allegedly walked off the job when at lunch time John presented him with a "meal" that consisted of nothing but a handful of walnuts. In food, as was the case with clothing, he appeared willing to subsist on other's castoffs. One Mansfield area farmwife opened her door one day to find John fishing scraps of stale bread out of a bucket of slop she intended to feed to her hogs. John responded to the startled woman by scolding her for wasting food perfectly suitable for humans on swine.[54] Yet Chapman may have only been passing along the advice of Lydia Maria Child, whose best-selling book, *The American Frugal Housewife*, advised women to "look frequently to the pails, to see that nothing is thrown to the pigs which should have been in the grease-pot. Look to the grease-pot, and see that nothing is there which might have served to nourish your own family, or a poorer one."[55]

Some who knew Chapman in his later years claimed he was a strict vegetarian and that his refusal to eat animal flesh was motivated by a desire to never harm one of God's creatures. There are many tall tales that recount his extraordinary aversion to harming any creature—his great remorse for striking a serpent that had bit him was just one. Another told that one cold night while sleeping out by a fire, he became distressed when he realized that mosquitoes attracted to the light were flying into the flames. John allegedly put the fire out and slept in the cold rather than contribute to the demise of some of God's tiniest and peskiest creatures. "He never carried a gun, never killed any game," R. I. Curtis recalled, but as we have seen, some evidence contradicts this.[56] Still, the claims of vegetarianism are frequent enough to merit consideration, and Chapman may have adopted vegetarianism later in life. If he did, he was not inspired to do so by New Church writings but may have been influenced by that rival group of Swedenborgians, the Bible Christians. Metcalfe's sermon on the abstinence from eating animal flesh was in circulation by 1821. That John, a voracious reader of all things Swedenborgian, may have encountered it and been influenced by it, is a reasonable possibility.[57] Sylvester W. Graham, a Presbyterian minister and America's most famous vegetarian, began preaching against meat-eating in the 1830s, and Yankee Bronson Alcott, discussed further in chapter 5, embraced vegetarianism because he believed man could not achieve spiritual perfection as long as he consumed the flesh of slaughtered animals.[58] That John left no written record of his philosophy of living, but instead a rich oral tradition that recounts his peculiar ways, makes it more difficult to sort out what is real and what is comical invention in the Appleseed stories.

John's determination to live a life of radical self-denial was in many ways just a solitary manifestation of a broader social reaction during the age when the market was transforming American life. The retreat into simplicity and emphasis on self-denial that was central to so many religious movements of the age was in large part an anxious reaction to the revolutionary impact that expanding capitalist markets were having on traditional values and ways. To be sure, many Americans embraced the promise of material prosperity capitalism brought. Frenchman Alexis de Tocqueville, who toured the United States in 1830 and 1831, believed that Americans were the most materialistic people in the world. Most of the time, Tocqueville insisted, "the desire to acquire the good things of this world is the dominant passion among Americans." But, he continued, "there are momentary respites when their souls seem suddenly to rush impetuously heavenward." Tocqueville believed that the Americans' extreme materialism and their uncontrolled manifestations of spiritualism were opposite sides of the same coin. He was fascinated with the phenomenon of the religious camp meeting, when "whole families, old men, women, and children, cross difficult country and come great distances" to listen to "preachers hawking the word of God from place to place." During these events, which often went on for four or five days, these otherwise materially focused farmers "neglect their affairs and even forget the most pressing needs of the body." Furthermore, many of them are open to considering unorthodox views which would have a harder time gaining traction in Europe. "Strange sects strive to open extraordinary roads to eternal happiness," he declared. "Forms of religious madness are very common there."[59]

American Methodists, who found early success evangelizing to the poorest and lowliest in society, often preached a message that encouraged the pious to resist the indulgences and luxuries of the market economy. Their emphasis on self-control and rejection of self-indulgence was relatively modest, however, and by encouraging hard work and thrift, it had the ironic effect of promoting a lifestyle in the new capitalist economy that often resulted in material prosperity. Francis Asbury, the leader of American Methodism, often worried that emerging prosperity of many Methodist adherents actually threatened to undermine their piety and zeal.[60] Yet other religious and utopian groups encouraged a more radical reaction to the materialism of the market. In the big cities of the East, some fairly well-to-do middle-class evangelicals embraced "retrenchment," replacing their fine clothes with plain ones and committing to a very basic diet consumed on primitive tables and with primitive utensils more commonly found on the frontier than in middle-class Boston and New York neighborhoods.[61] Shaker communalism and the complete rejection of private property was an even more dramatic manifestation of this trend, as was the emergence of groups preaching various forms of abstinence from alcohol, animal flesh, and sex. Bronson Alcott's short-lived Fruitlands community, discussed in the next chapter, was largely defined by the long list of pleasures and temptations its participants denied themselves, including meat, alcohol, tea, coffee, sugar, leather, wool, and cotton. John Chapman, in his radical commitment to living rough, to celibacy, to wearing the discards of others, and to subsisting on the simplest (and possibly vegetarian) diet was an extreme manifestation of this larger trend. But he appeared to be more successful than most of his fellow reformers in maintaining it. John's "primitive Christianity" came not from Swedenborg and the New Church but from a host of other religious and utopian ideas reacting to the changes that expanded markets were bringing to the nation. This rejection of the material temptation of the marketplace would continue to have appeal to the contrary-minded, including men like Henry David Thoreau, who retreated, if only a mile from civilization and only for a year, to a primitive cabin in the woods at Walden Pond.

The tension between the new materialism and piety resonates through one of the most frequently retold Johnny Appleseed stories. One warm evening in Mansfield, Ohio, Chapman was among the crowd listening to a traveling preacher expound from the town square. The theme of the sermon was the threat materialism posed to authentic piety. The preacher scolded his audience for endangering their souls by indulging in the luxuries that the expanding market economy had recently brought to Mansfield, like machine-made calico

"HERE'S YOUR PRIMITIVE CHRISTIAN."

The story of a ragged John Chapman's encounter with a traveling minister preaching the virtues of the primitive Christian was frequently retold across central Ohio, revealing contemporary obsessions about the relationship between materialism and Christian piety in an age of expanding markets. *Harper's New Monthly Magazine,* Nov. 1871, 836.

from Manchester and store-bought tea. At some point during this harangue, the preacher thundered rhetorically, "Where is the primitive Christian, clad in coarse raiment, walking barefoot to Jerusalem?" In response John Chapman, barefoot and wearing a coarse tow linen garment or perhaps even a coffee sack, walked up to the stump from which the minister was preaching and placed one of his black, horny feet on it, declaring "Here he is!"

The story was repeated frequently throughout central Ohio and even made its way to northwestern Indiana. While the details and coloration were altered slightly in each version, its popularity suggests that it was a joke that resonated deeply with a people who had experienced a rather dramatic transformation in their lifestyles in just ten years. Improved transportation and an influx of capital had lifted the more successful residents of north-central Ohio out of the barefoot, homemade cider and log cabin days into a more comfortable existence. Most welcomed these changes enthusiastically, but with all change came reflection on what had been lost. Preachers, especially, worried that the new comforts threatened to undermine the spiritual convictions of their flocks, and returned to the theme with a tiresome regularity. John Chapman's appearance at the minister's stump, a walking manifestation of a rejection of materialism that the minister could not match, no doubt gave them a brief and welcome respite from the guilt they were made to feel.[62] From such stories legends are born, and the story of Johnny Appleseed the primitive Christian, a symbol of resistance to sinful materialism, had the unfortunate effect of freezing in place a static vision of the man John Chapman, who contained as many contradictions as any other human and whose goals and aspirations changed throughout his life.

CHAPTER FIVE

To Serve God or Mammon?

In the decades after the War of 1812 the primitive central Ohio communities where John Chapman was building a life were rapidly transformed by the penetration of national markets into the hinterland. It was in these years that Johnny Appleseed the folk legend began to emerge. And one of the dominant threads woven through Appleseed stories was his radical commitment to a life of self-denial in an increasingly materialist age. Clothing himself in the cast-off rags of those around him who had since adopted finer accoutrements, preferring to go barefoot even after shoes were abundant and cheap, subsisting on the meanest fare, and getting by without a permanent shelter, John Chapman appeared to be living his life in opposition to the new materialism of his age. In 1810, his rough appearance and primitive lifestyle might have been understood as a result of personal poverty and the general scarcity of frontier life, but by 1830 it surely struck those who knew him as a deliberate choice. Poverty certainly had not been banished from central Ohio by that latter date, but it was a condition most commonly reserved for widows and orphans with limited means to support themselves, the physically disabled or mentally feeble, the drunk and dissipated. But John was none of these. He was healthy and strong and capable of doing a hard day's work. He had a vocation as an apple tree planter, and there was plenty of demand for his trees. He was not only literate but possessed a mind sharp enough to debate the finer points of theology. A few called him lazy. Some said he was mad. But most simply described him as peculiar. As discussed in the previous chapter, many concluded his lifestyle was inspired by religious conviction. John Chapman, most who knew him believed, *chose* a life of poverty in an age where material comforts were increasingly available. That was what made him such a worthy subject for gossip and tale-telling.

But a closer look at the life of John Chapman suggests that there were moments when it was less clear that isolating himself from the market and placing society over self were his foremost objectives. In his early years on the Pennsylvania and Ohio frontiers, John Chapman's austere lifestyle was more the product of necessity than choice, and there is evidence to suggest his aspirations were not much different from those of his frontier neighbors. He lived and dressed rough and subsisted on the meanest fare because that is what poor young men did on the frontier. In addition to his seedling apple tree business, John sometimes worked as a hired hand, as he had for Caleb Palmer before the war. Labor was scarce and typically commanded fifty cents a day. While not a princely sum, and often paid in goods rather than cash, as a mean-living bachelor, Chapman could not help but accumulate a modest amount of capital through day work.[1] Even before the war, he had managed to accumulate enough cash to pay fifty dollars for two riverside town lots in Mt. Vernon. The war brought temporary chaos to north-central Ohio, but with Perry's victory on Lake Erie, the threat to the region subsided, and a new prosperity returned to the region. The war itself had raised prices for farmer's goods, as armies needed to be fed and supplied. And with peace came renewed migration from the East, new families who needed all kinds of things, including apple trees, to start their new lives. If John had managed to accumulate some capital before the war through a simple strategy of working as a hired laborer and planting and selling apple seedlings, he no doubt accumulated even more in the postwar years without additional effort. This prosperity provided John with new opportunities and gave him new choices.

The opportunities created by the new market economy and the response of Americans to these new realities exposed what historian Stewart Davenport has called "one of the central paradoxes of American history: Americans are and always have been some of the most voluntarily religious people in the world as well as some of the most grossly materialistic."[2] Tocqueville recognized this contradictory impulse, but perhaps he did not quite understand the extent to which individual American Christians grappled to reconcile Matthew and Luke's warning that "ye cannot serve God and mammon" with Adam Smith's insistence that it was self-interest, not selfless benevolence, men were obliged to pursue.[3] "It is not from the benevolence of the butcher, the brewer, or the baker that we expect our dinner," Smith declared, "but from their regard to their own interest."[4] No doubt apple tree peddlers also struggled with the question of whom they should serve in this new age.

AN ECONOMY OF OWLS

Probably nothing transformed life along the Mohican and its tributaries between 1813 and 1818 as much as the mass infusion of paper currency into the local economy. From the outset, improvement-minded Ohioans had recognized that the scarcity of a medium of exchange was a hindrance to economic growth. By 1807, the state legislature began issuing charters to banks to address this problem. When Congress allowed the charter of the First Bank of the United States to expire in 1811, legislatures in Ohio and other states responded by chartering more banks, and some communities even formed unchartered local banks.[5] The appearance of state banks in central Ohio had revolutionary implications for the economy and for the way in which neighbors engaged in business transactions. When paper money was absent and coin scarce, most trade between neighbors had been straight barter transactions, with goods and services exchanged at the time of the deal or through promises to pay communicated orally or written out on paper. Written personal IOUs were transferable, as evidenced by John Chapman's 1818 note in which he promised to pay to Eben Rice "or bearer" eighteen apple trees.[6] Rice might have exchanged that note for other goods and services with a different neighbor but only if that other neighbor had confidence in the worthiness of John Chapman's promise. John's promissory note's value as currency was directly linked to his reputation. The farther afield the note migrated, the less likely it was to be accepted as payment. The ledger book method of extending credit, employed by country merchants, also depended on personal knowledge of and confidence in the customer. The ledger from John Daniels's store at Brokenstraw revealed that he allowed most of his white customers to settle their debts at a future date but that most transactions recorded with his Indian customers involved same-day exchange of goods for goods.[7] The handwritten IOU and the country store ledger might lubricate the wheels of local trade to some extent, but they had limited usefulness in advancing economic exchange to wider networks.

Specie—gold and silver coin—offered a form of exchange that freed buyers and sellers from dependence on personal connections. Americans understood silver and gold to have an inherent value that transcended the reputation of the holder, although incidents of counterfeit coin did occur. Except in the cases when the authenticity of the metal was in doubt, the value of specie was separated from the reputation of the bearer, and even from people's confidence in the institution which issued it. Spanish dollars made of gold and silver were as likely to be exchanged as coin emanating from the U.S. mint in Philadelphia. The security of its value was tied to the metal itself rather than to the issuer. Specie allowed perfect strangers to do business with each other. But it was in short supply on the Ohio frontier and could not adequately meet the needs of farmers in their business transactions. Paper money was meant to address this problem, as it amounted to a promise to exchange the note for specie by the issuing bank to the bearer. The credibility of the note now rested not on the reputation of one person but on the issuing institution. As a result, a local bank in good standing might expect its notes to circulate at face value locally, but as the notes made their way to more distant communities, they might only be accepted at a discount or could be rejected altogether. The notes of local banks depended on the collective reputation of their local directors and shareholders.

The village of Mt. Vernon, where John owned two town lots, was among the first in central Ohio to launch a banking venture. When their petition to the Ohio legislature for a state bank charter was denied, they decided to go it alone and launched the Owl Creek Bank as an unchartered venture. They capitalized it at $150,000, selling shares at $50 each but only requiring shareholders to put a small amount down and to promise to pay the rest on installment. The founders built a small but sturdy structure on Mt. Vernon's emerging Main Street. It was only about fourteen feet square inside, but to give the institution an air of substance, they adorned the front stoop with Doric columns and painted it red. Not scrimping on security, the investors used fourpenny nails in the window and door shutters to protect the contents from thieves. The bank board designed a series of attractive notes in denominations from twelve and a half cents to ten dollars, with a striking picture of an owl perched in a tree in the foreground and a water-powered mill, a symbol of the town's emerging industry, in the background. The Owl Creek Bank was launched with very little specie or chartered currency on hand. The value of the paper "owls" issued was dependent entirely on the promise of the bank's shareholders. Nevertheless, by 1816, paper owls were among the most common notes circulating across central Ohio.[8]

The proliferation of paper notes in the upper reaches of the Muskingum watershed brought rising prices and general prosperity and, perhaps unsurprisingly, the go-ahead spirit began to infect some settlers who had until this time appeared content to continue to pursue a self-provisioning lifestyle. What is more surprising is that John Chapman appears to have been drawn into the enthusiasm as well. In late May 1814 after a government land office opened in the new town of Mansfield on the Rocky Fork of Mohekan John's Creek, John entered an agreement with a Ms. Jane Cunningham to acquire a ninety-nine-year

Three-dollar bill issued by the unchartered Owl Creek Bank, Mt. Vernon, Ohio. After the War of 1812, "owls" and other notes with questionable backing proliferated across Ohio during the land speculation boom and contributed to the Panic of 1819. Author's collection.

---

lease on a quarter section of Virginia Military District school lands just southeast of town.[9] What John's relationship to Ms. Jane Cunningham was is not known. Ninety-nine-year leases on school lands offered a poor settler one of the best opportunities to acquire a farm legally. The Northwest Ordinance had established a plan for funding public schools that required one section of land in each township to be reserved for that purpose. But the Virginia Military District's settler-speculators—men of some wealth like Thomas Worthington—bought and carved up most of that slice of southern Ohio before Congress could apply its grid survey to the land and implement the school land reserve. As a result, Virginia Military District school land sections were carved out of the U.S. military district in north-central Ohio. The land could not be sold outright but was offered in quarter sections—160 acres—on long-term leases, renewable forever.

John Chapman and Jane Cunningham were able to make a claim for an up-front payment of just ten dollars. They would then have three years to build a good cabin and clear at least three acres for farming, a provision inserted to keep these lands out of the hands of absentee speculators. No additional payments were due for five years, at which time the lease holders were expected to make annual interest payments on the value of the land at 6 percent. Assessed at a value of two dollars an acre, the annual "rent" on Chapman and Cunningham's quarter section would be $19.20. If the payment were split equally, John might have raised his portion by working as a hired hand for a month or by selling at least two hundred seedling apple trees.[10]

Dividing a quarter section of land with a partner was a modest ambition for Chapman. The eighty acres he presumably controlled would be more than he needed if he chose to settle down into a life of farming. The typical farm family in these years might succeed in putting forty acres of land to good use, perhaps more with hired help. Even though Chapman now controlled more land than he could put under the plough, he caught the land speculation fever and was ambitious to acquire more.[11] In August 1814 he acquired another quarter section of school lands for ten dollars in fees just to the northeast of the town of Mansfield. The following February, he paid one hundred dollars cash for the rights to Richard Whaley's school land claim near the town of Wooster in Wayne County. Clearly, he expected the value of these leases to continue to rise. Then in April he acquired a fourth quarter section of school lands near Perrysville on the Black Fork. By the spring of 1815, John claimed three quarter sections of land outright and another in partnership.[12]

To secure these claims he needed to build four comfortable cabins and clear at least twelve acres in different parts of the region, no simple task for a single man. John needed to secure help constructing the cabins, which involved either payment in cash or apple trees or a reciprocal obligation to return the favor. Clearing land was a labor-intensive process. If John employed the method of his fellow New Englanders, he would have to chop down trees, root out the stumps, and burn the debris before plowing and planting. Although he had earned a reputation in his twenties for his ability to fell timber, he was in 1815 over forty years old—still fit by all accounts, but undoubtedly a bit slower. The New England method was rarely undertaken by a family alone and instead depended on the help of neighbors who came together for a "logging bee." Alternatively, Chapman might have adopted the simpler method employed by his Southern neighbors: girdling the trees and allowing them to rot slowly, while growing corn amid the dead giants. Dead trees were only removed or burnt as they were toppled by storms. If the New England method demanded more labor at the outset, the southern method still required a significant expenditure of labor for clearing, spread over several years. And neither method resulted in rapid deforestation. Ohio agricultural historian Robert Jones has suggested that a typical farm family could find time to clear only about two to three acres of land a year in between other necessary tasks.[13] While it was common for frontier neighbors to help one another in the most labor-intensive tasks, accepting the help of neighbors brought on reciprocal expectations. If John intended to fulfill the claim requirements on each of these quarter sections within three years, he was committing himself to a substantial amount of work.

Of course, it is also possible that John had no intention of fulfilling the requirements when he made these claims but instead, caught up in the speculative fever of the period, intended to resell them to one of the many newcomers arriving each spring before the three-year time limit had expired. Whatever his intentions, this turn toward speculation is not in keeping with the image of Chapman as the primitive Christian, living on just what he needed and no more. It is difficult to explain John's land fever without concluding that he, like many of his neighbors, had been drawn into the possibilities of wealth that this postwar boom was offering and was ambitious to make money. Parcels of land as large as 160 acres were hardly useful to him in his nursery business, which required small plots close to water sources.

Chapman sat tight on his small empire for several years, as more migrants flooded into the region, more paper currency flowed through the economy, and the prices of everything, including land, rose. But this bubble, fueled in large part by a storm of paper with questionable backing, would eventually burst. The signs of trouble were already present by the beginning of 1817. Many in the state expressed concern about the proliferation of unchartered banks like the Owl Creek Bank of Mt. Vernon. One Columbus newspaper recounted a story that suggested the worst insult that could be levied upon another person at this time was to call them a "d——d UNCHARTERED son of a B——h."[14] The Owl Creek Bank directors had issued paper on the promises of shareholders who had to that point only paid a fraction of the value of their share. The bank made several requests to shareholders for additional payments late in 1816, but many were delinquent. Nevertheless, people continued to circulate owls in Mt. Vernon and in other towns across the state.[15]

When, in the fall of 1817, a traveling Yankee peddler named Silas Chesebrough arrived in Ohio, he had some difficulty selling his wares, even though the items he had brought from Connecticut were still scarce in Ohio. "There is here so much bad money or money that is good for nothing out of its own neighborhood, that it is hard work to sell goods," he declared.[16] When he arrived at Perrysville on the Black Fork, he noted that the residents had plenty of hogs, cattle, and wild honey, but "Dry goods and Groceries are very high, and as for money, I may truly say they have none."[17] It was more likely that they simply did not have the kind of money Chesebrough was willing to accept. In Mt. Vernon, Chesebrough visited the Owl Creek Bank to exchange some of the notes he had accepted from customers along the way, but the bank refused to exchange its own notes, either for specie or for notes from more reputable banks. Chesebrough had no choice but to exchange them for goods from a local store at cost.[18] Despite the problems of unreliable money, Chesebrough found that speculation in land was unabated, and lands sold by the government just a few years earlier for two dollars an acre were in some places now fetching twenty dollars.[19]

By December 1817, reports that the Owl Creek Bank was not honoring its own notes had spread beyond Mt. Vernon to other towns in the region, and antibank feeling grew. One angry resident appeared in the bank one day with a large owl he had killed, dropped it on the bank counter, and declared "I have killed your bank President!"[20] On January 1, 1818, the directors of the Owl Creek Bank published a notice for a meeting of all shareholders, two months hence, to discuss closing the operation down. But, as many shareholders had defaulted on their payments to the bank, Owl Creek's directors were not able to honor all of its obligations.[21] The settlement of claims against the Owl Creek Bank tied up courts for much of the next decade. The scene in Mt. Vernon was being repeated throughout the West in 1818, where financial troubles were well under way before the Panic of 1819 occurred nationwide. In such a climate, many local speculators began to retrench, but John Chapman, apparently oblivious to the impending crisis, extended his obligations by acquiring a fifth quarter section of school lands from William Huff for an undisclosed sum.[22] Interest payments on his four and a half quarter sections of land would soon come due. Throughout the summer and fall of 1818 John bought and sold land in a pattern that leaves little confidence in his business acumen. In June he sold off several parcels of his interest in the quarter section he owned closest to the village of Mansfield, a move that made sense because it would lessen his impending commitment to interest payments. But in August he paid $120 cash for a town lot in Mansfield, which he sold at a twenty-dollar loss less than three months later. In November he sold another twenty-acre parcel of his Madison township quarter section to Mathias Day, leaving him with just 70 of the original 160 acres. In 1820 he made his first and only interest payment on the first quarter section he had acquired with Jane Cunningham. In the ensuing depression years, John defaulted on that land and three other quarter sections, managing to keep up payments only on the seventy acres he retained near Mansfield.[23]

Hard times lingered in the region until at least 1823, when some signs of recovery began to show. The bust following the boom no doubt caused many settlers in the region to retreat by necessity into a self-provisioning lifestyle and at least momentarily to reconsider the wisdom of their venture into the marketplace. John Chapman likely did so as well. The postwar boom and bust may have proven to be defining moments in Chapman's life. With the return of prosperity to the region after 1823, he did not return to his speculative ways.

John's flirtation with, and retreat from, the market economy was not unique in antebellum America. It is instructive to revisit the story of another Yankee dreamer who later in life strove to embrace primitive simplicity, the utopian vegetarian Bronson Alcott. While Chapman was twenty years older than Alcott, both men arose in similar circumstances, and they appeared to have been shaped by similar forces. Like Chapman, Alcott grew up on an overcrowded Connecticut Valley farm. While Alcott's father owned the property, he was saddled with so much debt that his hold on the land was at times tenuous. Like thousands of other young New England men in the first decades of the nineteenth century who saw limited opportunities in the old towns that raised them, Alcott set off to make his fortune as a traveling Yankee peddler.[24] Alcott, like Chapman, was the eldest son in a large family and felt the push to move on at a young age. In 1815, at the age of sixteen, Alcott left his Connecticut home and took up peddling after failing to find a job as a teacher. While Chapman was selling apple trees in central Ohio, Alcott was hawking almanacs across Virginia and the Carolinas.

The term "Yankee peddler" carried a certain taint of immorality in an age when some were still ambivalent about markets and cash economies. Chesebrough and Alcott made money by buying goods at a low price in one location and selling them at a high price in another. Furthermore, while farmers produced wealth from the soil and their labor, peddlers appeared to produce wealth from nothing. At a time when notions of a just and fixed price for goods still prevailed in the minds of many Americans, peddlers operated outside of the bounds of this old notion.[25] Yankees who came west to peddle goods or build towns earned a reputation as shrewd and slightly shady men on the make, a stereotype that became so powerful in the West it was imposed on virtually all Yankee migrants. One observer declared that "the whole race of Yankee peddlers . . . go forth annually in the thousands to lie, cog, cheat, swindle."[26] Even John Chapman, who became famous for his generosity and his apparent lack of business savvy, could not entirely escape this stereotype, if the Tannehill family tradition that Melzar didn't want to commit his daughter to marriage to a "pedlar" is to be believed.[27]

Over the next several years, Alcott made five trips south peddling almanacs and other goods, and early success seemed to have awakened a materialist spirit in him and pumped up his pride. He became confident that he would be able to quickly accumulate enough wealth to pay off his family's debts, telling his father that he "would have them out of debt by his twenty-first birthday."[28] But his confidence fell as his profits dissolved. In the wake of the Panic of 1819, prices for goods began to free-fall as demand dried up. The inexperienced and overconfident Alcott soon found his debts mounting. By 1822, he could not pay the merchants whose stock he had acquired on credit, even by returning his extra merchandise and handing over his horse and wagon. Rather than saving his family from financial trouble, he turned to them for help. By 1823, Alcott was $600 in debt to his father.[29]

Alcott's experience as a peddler appeared to sour him, temporarily at least, on markets and materialism. In March he wrote to his cousin William about his peddling experience. "I am [not] disposed in this respect to accept the saying, sometimes quoted by peddlers, that 'Peddling is a hard place to serve God, but a capital one to serve Mammon.' I find I have not served either to the best advantage, and wish I may find the grace to amend my ways. I have enjoyed some of the pleasures and profits of travelling, along these sandy roads, and in the society of this simple people, living here beside their juniper forests, in the midst of their pastures and fields, their flocks and herds, and their old orchard."[30] Alcott may have found new admiration for the self-provisioning lifestyle of the "simple people" he encountered as a peddler, but he did not join them in their lifestyle. He took a job as schoolmaster, eventually married, and began remaking himself as an educational reformer. Bronson Alcott embraced, in principle at least, a life of simplicity and self-denial and found a partner who was willing to follow him through a life of economic insecurity. Abba May was deeply attracted to Bronson Alcott and did not need to be persuaded by her brother who told her, "Don't distress yourself about his poverty. His mind and heart are so much occupied with other things that poverty and riches do not seem to concern him."[31] The description might have been written about John Chapman as well.

Alcott's story is told in his own writings and in those of the literate company he kept. It is easy to see its twists and turns. Alcott's youthful embrace of the market and materialism and his disillusionment with it in failure, his practiced efforts to live simply, and the regular temptations to acquire material things that sent him repeatedly into economic misfortune are well documented. By contrast, most details of Chapman's life escape us. It is worth considering whether John Chapman faced similar temptations throughout his life. Surely his efforts to secure control of hundreds of acres of land before 1819 were inspired by a hope that at some later date he could sell these claims at a handsome profit. And it is also worth reflecting on the reason Melzar Tannehill gave for not wanting his young daughter to marry John Chapman: that he was a "pedlar," a term loaded with perceptions of unscrupulous material ambition.[32] Was there a moment when others in central Ohio might have perceived him this way as well?

John only imperfectly fit the model of the Yankee peddler in these years. Most were single men who traveled from the East with a wagonload of clocks, books, or other items scarce on the frontier, seeking to make a profit as they wandered from village to village. Perhaps the association came in part from his last name, "Chapman," which was a synonym for peddler in early America. Many states began passing laws regulating the activities of "hawkers, peddlers and petty chapmen" in the late eighteenth and early nineteenth century.[33] Furthermore, most traveling peddlers were from New England, so John's Yankee origins, his status as a single man, and his general itinerancy connected him to the powerful, and negative, stereotype of the Yankee peddler. Furthermore, John was in fact peddling a product—the seedling apple tree—to rural customers. He was also in his way "hawking" books, as he traveled from cabin to cabin, seeking to interest families in his Swedenborgian literature. That he asked no price for them would not have immediately set him apart from other Yankee peddlers, who were famous for their crafty sales tactics. Yankees peddling shelf clocks on the frontier did not immediately introduce themselves as salesmen. They sought to gain the confidence and trust of their targets at first then might ask to temporarily "leave" a beautiful shelf clock with a country woman on the pretense that it had been purchased by a neighbor. The peddler understood that she was eager to lift her family from their rude frontier existence into middle-class respectability and did not want to fall behind her neighbor in this pursuit. Once the clock had been placed on the mantle for safekeeping, she would have to have it, and she would convince her husband to pay up.[34] The calico fabric, the shelf clocks, the tinware, and the books peddlers brought to central Ohio cabins were not simply consumer goods—they were paths to respectability, and peddlers understood that the person in the family most desirous of that status was most often the frontier wife.[35] John Chapman's seedling trees did not carry the same associations, but they could be a step up for a poor frontier family with ambition to have that seed-grown orchard top-grafted with respectable stock at a later date. And despite his rough and rugged appearance, for many of the frontier women who welcomed him into their homes, he was also peddling to them a kind of refinement still scarce in Ohio. Rosella Rice's childhood memory of Chapman's eloquence in describing apples illustrates this attraction. "His description was poetical, the language remarkably well-chosen; it could have been no finer had the whole of Webster's Unabridged, with all its royal vocabulary, been fresh upon his ready tongue. I stood back of my mother's chair, amazed, delighted, bewildered, and vaguely realizing the wonderful powers of true oratory. I felt more than I understood."[36] That the women of the household were generally more eager to see him coming than their husbands also mirrors the often gender-diverging responses to the appearance of a more conventional Yankee peddler at the cabin door.

But if his origins and circumstances might have led some people to view John Chapman as a Yankee peddler, he did not appear to possess the characteristic shrewdness of that archetype. By 1823 he had forfeited all but one of his Virginia Military District school land leases, retaining control of just 70 of the original 160 acres. This land, in Madison township, must have meant more to John than any of the other parcels he had acquired during his speculative frenzy, as he kept up the interest payments on it for the rest of his life. Whether he built a cabin there is unknown. It is possible he kept it for the use of his younger half-sister Percis and her husband William Broom.[37] The Brooms joined Chapman in the Mansfield region some time in the late 1810s or early 1820s and were reportedly quite poor, depending on John for support. Rosella Rice contrasted John's eloquence with the plain-spoken language of his sister Percis, who, Rosella declared, "Was not at all like him; a very ordinary woman, talkative, and free in her frequent, 'says she's' and 'says I's.'"[38] The appearance of Percis in the region also puts to lie the story that John Chapman was a loner without family. Although he did not make his life near Marietta, where the majority of his half-siblings were, John kept family connections open throughout his life.

The few clues in the historical record related to Chapman's activities once the financial-panic-induced depression had lifted suggest that he struggled to adjust to the new economic and social realities. As more migrants flooded into central Ohio and became legal property owners, his claims to his apple seed nurseries, which were scattered across lands he did not legally own, began to come under assault. In 1823 Chapman paid forty dollars to Alexander Finley for a 2.5-acre flood-prone parcel along a branch of the Mohekan. Just two years later, he acquired another 14.5 acres of low-lying land just a few miles from the Finley parcel from Isaac and Minerva Hatch for sixty dollars "lawful money of the United States to us in hand." This low-lying parcel, today part of the Funk Bottoms Wildlife Area, is now permanent wetland and a popular spot for bird-watching. It may not have been permanently inundated before the state built a series of flood-control dams in the 1930s, but it was probably subject to almost annual seasonal flooding in Chapman's time, as was much of the land in this region. In both cases, John appears to have paid cash up front for the parcels, and both were relatively small acquisitions—too small for setting up a home and farm but plenty large enough to shelter seedling nurseries. Both had easy access to the water necessary for sustaining young trees, something that was both an asset and a curse. John's hold on both parcels was also short lived. He apparently never paid taxes on the Finley lot and abandoned it. The 14.5 acres he acquired for sixty dollars in 1826 he sold for just fifty dollars six years later, when good land should have increased in value. It appears that he paid more than the value of these lands in both cases, probably by necessity. He may have "squatted" these lands years earlier, planted valuable nurseries on them, and then found himself forced to pay the landowner's asking price just to get access to these trees at the moment when they were most suited for selling.[39]

That Chapman was increasingly having trouble accessing his squatter nurseries as central Ohio filled up is also supported by a story recalled by John James, a fellow Swedenborgian who first encountered Chapman in 1826. James had come to Swedenborgian convictions at first through marriage, having married the daughter of Francis Bailey, the Philadelphia printer who had produced the first American editions of some of Swedenborg's writings. According to James, Chapman carried a letter of introduction from New Churchman Alexander Kinmont, who had come to Cincinnati to establish an experimental school. He also made it clear to James that he was well acquainted with his wife's family. The purpose of Chapman's visit was to enlist James's help in getting access to a nursery he had planted on nearby land since acquired by a new owner, who had no knowledge of the informal agreement Chapman had reached with the previous property owner. He knew James was a man of significant standing in Urbana, and hoped that James might be able to facilitate an introduction to and agreement with the new landowner. According to James, Chapman was not too anxious about the situation and was prepared to give up on the claim if it was too much trouble. Still, the trees were important enough to Chapman for him to have gone to some trouble, in the hope that it would help him regain access.[40]

The story reveals a great deal about the rapidly changing nature of central Ohio society. By 1826, central Ohioans were increasingly committed to relying on formal instruments for affirming all economic agreements. That John lacked any written record of his agreement with the previous landowner was enough to empower the new land owner to deny him access to his trees, even though that owner knew he had not planted those trees himself. Chapman was a stranger to this landowner, whom he felt free to disbelieve. His response was to draw on his web of personal connections to demonstrate his trustworthiness, a strategy that may have worked for him in the past. Whether Chapman's efforts succeeded in this case we do not know. James does not mention in his account whether he was successful in helping Chapman regain access to his trees.

Taken together, the two documents recording John's 1823 land acquisitions and the story recounted by John James reveal one of the significant ways Ohio was changing in the mid-1820s, as the population increased and open land became scarce. The squatter strategies and informal agreements that prevailed before the War of 1812 could not be counted upon in this new, more crowded world, filling up with newcomers and strangers. In order to sustain his nursery business, he would have to secure the rights to plant land through purchase, lease, or written agreements.

One last surviving document from 1826 supports the theory that it may have been the year when John Chapman began to adapt to the new rules. In that year he leased a half acre in Ashland County's Green township "where the said Chapman plants fruit" for a forty-year term and the price of just twenty apple trees. That agreement was formalized and recorded in the Richland County deed records a few years later. As markets and formal instruments replaced personal connections in the central Ohio economy, John Chapman, by necessity, adapted, and in subsequent years he recorded more formal lease arrangements at county land offices. But he did not adjust to the new materialism that came with this transition. In the final two decades of his life, John increasingly stood out and attracted attention because of his bold rejection of the new materialism.

AN AGE OF IMPROVEMENT

By the middle of the 1820s, the go-ahead spirit had returned to the nation and to Ohio. "Improvement" was the watchword of the day. "'Improvement' in its early nineteenth century sense," historian Daniel Walker Howe explains, "constituted both an individual and collective responsibility, involving both the cultivation of personal faculties and the development of national resources."[41] Americans sought intellectual improvement through education, moral improvement through religion and a host of new moral reform organizations, and material improvement by applying up-to-date knowledge to their labor. Improving farmers formed agricultural and horticultural societies for advancing knowledge and employing the best practices. By doing so, their crops would increase in both quantity and quality and command higher prices at market. For the improving farmer, intellectual growth, personal discipline, and material wealth advanced together.

At the collective level, improvement meant using government, voluntary organizations, and economic cooperation to transform forest to field and knit together tidy communities with good roads, improved waterways, canals, and, later, railroads and telegraph lines. These collective improvements were necessary to enable the individual to reach his potential. This improving spirit had its most dramatic manifestation with the completion of the Erie Canal, a man-made river that seemed to defy the laws of nature, directing water to go where men needed it rather than where natural landscape features and gravity dictated. When completed in 1825, the Erie Canal established an all-water route from the Great Lakes to New York City.[42] Overland transportation, even on improved roads, had been a slow and costly way to move heavy, bulky farm goods from Western farms to Eastern cities. Now tons of goods floating upon water could be moved vast distances employing just the energy of a solitary mule. Ohio dairy farmers who lived within a day's wagon ride from Lake Erie could now sell cheese to customers in New York City. The success of the Erie Canal set off a canal-building mania, and Ohio soon launched its own plan for a series of canals to link Lake Erie to the Ohio River on both the eastern and western sides of the state. Once completed, the economic isolation that was in part responsible for the self-provisioning lifestyle of so many Ohio farmers would disappear.[43]

The movement for improving agriculture emerged first in the eastern states, where open land was scarce, population density was greater, urban markets were easily accessible, and the fertility of the land began facing sharp decline after generations of farming had depleted the soil. In the wake of the Panic of 1819, interest in agricultural improvement accelerated in the older states, and by the middle of the decade, about a half dozen Ohio counties had formed agricultural improvement societies. In these early years, interest in organizing to improve Ohio agriculture was mostly limited to the wealthiest Ohio farmers, who subscribed to Eastern agricultural journals to keep up with new developments. The *American Farmer* was launched in Baltimore in 1819 and within a few years had over 1,500 subscribers, including some in Ohio. By the 1830s, agricultural journals such as New York's *Genesee Farmer* had built national circulations more than ten times as large.[44] In Ohio, Cincinnati's first journal devoted exclusively to agriculture, *The Western Tiller*, appeared in 1826. Several more farm journals were launched from that city in the early 1830s, and Columbus saw publication of *The Ohio Farmer, and Western Horticulturist* in 1834. Editors of regional farm journals regularly recycled articles and letters that had appeared in Eastern journals, so new ideas in farming and horticulture were spread, if not adopted, quickly.[45]

Among the ideas most persistently championed by the early agricultural journals was the formation of local agricultural clubs and county agricultural societies.[46] Yankee migrants often led the way in this regard, and many of the first agricultural societies were launched in counties with a significant Yankee presence. Hamilton (Cincinnati), Washington (Marietta), and Trumbull (Western Reserve) all formed agricultural societies in 1818 or 1819. The mixed-origin central Ohio counties were a bit slower to start. Muskingum County (Zanesville) founded one in 1827, Richland (Mansfield) in 1829, and Wayne (Wooster) in 1833.[47] These new societies took their agendas from Eastern agricultural societies, so ideas that gained steam first on the East Coast soon found some advocates north of the Ohio River.

A persistent theme in the early agricultural journals was the reformer's frustration with simple country farmer's apparent indifference to the valuable knowledge the "book farmers" were offering them. The editors and contributors to these journals routinely attributed the slow progress in the adoption of agricultural techniques to character flaws of hinterland farmers. They were lazy and backward, skeptical of science, and ruled by tradition and superstition. Of course, the advocates of agricultural improvement generally commanded far more capital and labor, and had more freedom to take risks, than the poorer farmers who frustrated them. They could afford to invest in more expensive grafted trees and also could hire help to carry out their new practices, which were often more labor intensive. It is no surprise then, that John Chapman, who continued to peddle inferior seedling apple trees, came to be seen by some improvers as lazy, improvident, and driven by superstition rather than reason. Among the most oft-repeated stories about Chapman concerned the source of his aversion to grafting apple trees.

Some chroniclers wrongly suggested that Chapman's alleged aversion to grafting apple trees came from Swedenborgian beliefs. In fact, Swedenborg himself grafted apples in his own orchard, and never suggested that grafting violating God's laws. W. D. Haley's *Harper's Monthly Magazine* article, which appeared in 1871, helped to spread the notion that Chapman's aversion to grafting was based on irrational superstition by asserting that John "denounced as absolute wickedness all devices of pruning and grafting, and would speak of the act of cutting a tree as if it were cruelty inflicted upon a sentient being."[48] By the year the *Harper's* article was released, grafting was so universally embraced by orchard-keepers that many had forgotten why any rational person might choose to plant apples from seed. An account that appeared in a Jefferson County history about ten years after the *Harper's* account made the improbable claim that John Chapman abandoned that region because he was offended to learn that Jacob Nessly was grafting trees a few miles away on the Virginia side. But if this had in fact been the case, Chapman would not have settled anywhere in Ohio, as there were almost always grafted orchards within a day's walk of his seedling ones. The Jefferson County historian had clearly read the *Harper's Monthly* story and lifted some of its language in his own account, claiming that Chapman declared to his Jefferson County host, "They cannot improve the apple in that way—that is only the device of man, and it is wicked to cut up trees in that way... the correct method... is to select good seeds and plant them in good ground and God only can improve the apple."[49]

The most reliable explanation of John's aversion to grafting comes from John James's account of his 1826 encounter with Chapman. James claimed that Chapman told him that he did not graft but that "the proper and natural mode was to raise fruit trees from seed." James understood New Church teachings enough to know that they included no sanction against grafting, and while he noted that Chapman was emphatic on the matter, he did not impart any superstition to his preference for seedlings.[50] While folk tales that portray Chapman as harboring an extreme aversion to harming any living thing, including rattlesnakes and mosquitoes, might lend credence to the idea that John harbored some irrational aversion to cutting a tree, it is wise to treat that explanation with some skepticism, as it dovetails nicely with the perspective of the defensive "book-farming" agricultural improvers who dismissed poor, self-provisioning farmers' reluctance to embrace their new methods as a result of their ignorance and superstition. But John really required no external justification—scientific or superstitious—for his practice. There was good local demand for his seedling trees into the 1830s.

Yet by the end of the 1820s, the threat to John Chapman's seedling orchards was not simply economic and scientific but, increasingly, moral. Fruit from a seedling tree had many uses for a family of course, but in a world where more and more farmers grew food to sell rather than consume, the seedling orchard came to be seen as having just one purpose, the production of hard cider and cider brandy.[51] Per capita alcohol consumption rates in the United States were among the highest in the world in the 1820s, and a temperance movement emerged to combat the evil of drink. Whiskey and other distilled spirits were the focus of many of the early temperance advocates, as their extremely high alcohol content made them the greatest threat. Homemade apple cider began its life free from alcohol, then gradually fermented to a content of 6 to 8 percent. "New cider" was the term most frequently used for that beverage that had yet to ferment. When the circuit-riding minister Peter Cartwright stopped at an inn for a dinner of hot bread, he and his companion washed it down with "new cider." It was not until they mounted their horses and renewed their journey that they realized the refreshing beverage they had so eagerly quaffed was not so new, as both men had trouble staying upright on their horses.[52] Because the alcohol content of cider was not enough to ward off the growth of bacteria, cider-makers frequently fortified it with corn whiskey or other distilled spirits to get its alcohol content above 10 or 12 percent. Another common method was to leave the barrels outside on a cold night and scrape off the ice that formed in the top half of the barrel overnight; what was left behind, often called applejack, had roughly double the alcohol content.[53]

Some early advocates of temperance viewed hard cider as relatively harmless, even as a healthful alternative to distilled spirits. Benjamin Rush was among the earliest advocates of moderation in the consumption of alcohol and boasted that a crowd of seventeen thousand who had gathered in Philadelphia on July 4, 1788, celebrated with nothing but beer and cider, "those invaluable FEDERAL liquors" in Rush's construction, in contrast to hard liquor, which he associated with antifederalism. Rush also produced a "moral and physical thermometer" assessing various beverages on their health and moral properties. Water, milk, and "small beer" were at the top and the drinks of the truly temperate, while gin, whiskey, and rum were the beverages of the most dissipated. But Rush placed "Cider and Perry" just below the beverages of the truly temperate and suggested their effects were not all bad—that they produced "Cheerfulness, Strength, and Nourishment, when taken only at meals, and in moderate quantities."[54] Farm laborers seemed to agree, as farmers who took on hired hands in the harvesting season were expected to provide their workers with a ready supply of hard cider to refresh them and keep up their strength.

In its earliest years, leading advocates of temperance called for moderation in the consumption of alcohol and the avoidance of distilled spirits. But as the movement gained steam in the 1820s, an "ultraist" faction began pressing for total abstention from alcohol. They rejected the view that cider-drinking should be tolerated or seen as a "healthy" alternative to distilled spirits. A man could get drunk on cider as surely as he could on whiskey. "It takes a long time to make a man a drunkard on cider," one anti-cider crusader declared, "but when made, he is thoroughly made, is lazy, bloated, stupid, cross and ugly, wastes his estate, his character, and the happiness of his family."⁵⁵ Another ultra who campaigned against hard cider claimed its effect on families was often worse than distilled spirits. He cited the wife of a cider-drunkard who had told him that "cider made [the drunkard] more brutal and ferocious in his family. Rum overcame him quicker, laid him prostrate and helpless on the floor, or in the ditch; cider excited him and gave him the rage and the strength of a maniac."⁵⁶

By the late 1820s, the ascendant ultraist faction in the temperance movement began to point an accusing finger at the farmer's seedling apple trees. An article republished in religious and agricultural journals across the country in 1827 raised the question "What Shall I Do with My Apples?" Its author was determined to make the cider apple the newest front in the contest between God and Mammon, declaring that this was a question every Christian farmer should be asking himself. "If he gathers his apples, of course he must make them into cider; and if he makes the cider, of course he must sell it; and if he is to sell it, of course he must sell it to the distiller, or procure it distilled and then sell the brandy; and if the brandy is sold, it must be drank, and in this way every barrel will make and circulate liquid fire enough to ruin a soul, if not destroy a life." The author of the piece rejected the farmer's argument that to leave apples to rot was to waste God's bounty and instead cast the farmer as acting in pursuit of profits in the marketplace at the expense of his neighbors. "If no other market can be found for our cider, but at the still, let it be a matter of conscientious inquiry with every farmer, whether it is *right* for him to make more cider than he wants for reasonable use in his own family" when he knows his surplus will be distilled into cider brandy and sold to the intemperate. The author of the piece concluded that the only righteous path for the Christian farmer to take in regards to his seedling trees was to "BURN THEM."⁵⁷

Others began to pile on. A correspondent in Cincinnati's *Western Observer* said that farmers who raised fruit for the cider and brandy market would only join the temperance crusade "if ever the dictates of conscience get the ascendancy over the mammon of unrighteousness," framing the problem in the familiar language of a struggle between market-driven capitalism and Christian values.⁵⁸ A letter in *The Western Recorder* condemned "a certain innkeeper for purchasing two barrels of Hard cider for $1.25 each" and selling it to his customers, one of whom "had a bill of $8 charged to him for his portion of the cider, which, at six pence a quart, would amount to thirty-two gallons!" The innkeeper was "a professor of religion" and therefore should have refused to sell alcohol. Instead, he had made a nice profit at the drunkard's expense. By putting "the bottle to the mouth of his neighbor, [he] has done the deed, and is now calculating . . . on the wages of his iniquity."⁵⁹ *The Western Christian Advocate* recounted the story of a good temperance man and churchgoer who was regularly in the habit of turning his surplus apples into cider until the day he encountered a neighbor and fellow church member who got drunk on it. Troubled by his complicity in his friend's downfall, he destroyed his cider press and never made the beverage again.⁶⁰

But other campaigners against cider, perhaps conceding that in a contest between piety and profit the latter would ultimately win, began to make the case that taking apples to the cider mill was neither in the farmer's moral nor in his pecuniary interest. The costs of the hired hands required to gather all of the apples, the efforts to get the fruit to the cider mill, and the fees for processing them into cider were greater than the returns, these writers claimed, as cider fetched such low prices on the market. A more profitable strategy was for the farmer to let the hogs have those apples not fit for fresh eating. Self-provisioning farmers had long followed this practice, as it kept orchard floors clean and their small stock of half-wild pigs fat, and it required little of their scarce labor. But now horticultural improvers converted to temperance promoted fattening hogs on cider apples to a class of farmers concerned with maximizing the revenue on their farms. Those seedling apples lying on the orchard floor could be "converted" to pork with little effort. Both religious magazines and the agricultural journals devoted to improvement enthusiastically embraced the strategy of fattening hogs destined for markets on the old orchard's windfalls.⁶¹

By 1829, at least a few farmers had taken the advice of "BURN THEM" to heart. One report circulated in several journals told of a New Haven, Connecticut, gentleman who "ordered a fine apple orchard to be cut down, because the fruit may be converted into an article to promote intemperance." The editor of the *New York Enquirer* mocked this wasteful action, opining that "in this age of *Anti*-Societies, we may soon see the worthless of the land in league, to establish an *Anti-Apple and Anti-Rye* Society. Every time and age has its mania:—We hope the world will *sober* down before dooms-day."⁶² The story of the monomaniacal temperance man destroying apple orchards became part of the folklore of New England and the Midwest, and not only missing orchards but also abandoned ones were attributed to the zeal of the reformers.⁶³ Years later, Henry David Thoreau bemoaned the loss of seedling apple trees across the New England countryside. He repeated a story he had heard "of an orchard in a distant town on the side of a hill where the apples rolled down and lay four feet deep against a wall on the lower side, and this owner cut down for fear they should be made into cider." Thoreau was nostalgic for that time "when men both ate and drank apples, when the pomace heap was the only nursery, and trees cost nothing but the trouble of setting them out," a time long past when he penned these words in 1859.[64]

The number of orchards actually chopped down by temperance ultraists was likely not as great in reality as in local folklore. But the endless moral castigations orchard-owning farmers faced from the temperance crusaders seemed to have an effect. Many a pious farmer was surely troubled by the accusations that by selling apples to cider mills he was serving Mammon instead of God. Many writers attributed the abandonment and neglect of old seedling orchards across New England to the temperance crusade, their owners apparently deciding to forgo the attacks on their moral character by simply neglecting orchards and letting nature swallow them up.[65]

John Chapman left behind no surviving commentary on the temperance movement's crusade against seedling apple trees, nor do any of the early stories retold about him give us any clear signals as to where he stood on the issue of hard cider. Was John Chapman "on the side" of alcohol or "the side" of temperance? The question only makes sense to the modern student. It would have made little sense in Chapman's own time, when, as Thoreau noted, men "both ate and drank their apples." John did not arrive on a frontier filled with people who viewed alcohol as immoral and deliver to them the seeds of moral corruption. Alcohol was a staple item in the earliest days. If some of those frontier families eventually converted to teetotalism, they did so after central Ohio spring airs were already filled with the scent of the apple blossom. John himself was not a lifelong teetotaler. Holland Company store records from his days in Franklin, Pennsylvania, indicated that as a young man he purchased whiskey. Some central Ohio traditions passed down suggest he never touched alcohol, but John Dawson, who knew him during his Indiana years, noted that Chapman "was generally regarded as a temperate man . . . but occasionally he would take a dram of spirits to keep himself a little warm, as he said."[66] No surviving accounts portray him as drunk or a heavy drinker. He may have in his later years added alcohol to the long list of things he denied himself in his personal quest to live the life of the primitive Christian. He certainly did not abandon his lifelong campaign to plant seedling trees, even though he knew that many, perhaps most, would produce apples for the cider mills and brandy distilleries. That John struggled to live a life that served God, not Mammon, is in no doubt, but he apparently never concluded that doing so required him to abandon his vocation as "a gatherer and planter of apple seeds."[67]

While the temperance movement continued its crusade against seedling orchards, proponents of agricultural reform also continued their campaign to get farmers to replace old seedling orchards with grafted commercial varieties. If a farmer could not afford the time and expense of uprooting a seedling orchard and replacing it with nursery trees of good, grafted stock, then at least he could top-graft the old trees with promising new varieties. A frequent problem with this strategy was that it relied on the trustworthiness of the traveling top-grafters who performed this service for a fee and who would be long gone from the region before a farmer could know for certain that he had received top-quality grafts for his money. One county history claimed that many of the seedling trees had been top-grafted at an early date but nonetheless continued to produce low-quality fruit. Perhaps the farmers of the county had been swindled by a traveling grafter.[68]

Horticultural reformers advised their readers to plant several varieties: some early, some mid-season, and some late ripening trees. But they were especially enthusiastic about promoting "winter apples" as these had the most promise in the marketplace. In the midst of fall harvests, apples were so plentiful and their price so low that in some places they were hardly worth carting to markets. But "winter apples," which ripened late and had good keeping qualities so that they might be brought to markets in January, February, March, or later, could command much higher prices. Whether writing about converting apples to pork or the value of winter apples, the new horticultural journals encouraged farmers to stop seeing their apples as simply God's easy bounty, bestowed upon them in season to help sustain the family, and to begin thinking of them as important commodities.

Not only did the journals begin publishing the current prices of apples in all their forms—fresh, dried, and converted to cider—in specific markets, but they also began publishing articles about farmers who had received fantastic yields of quality apples from their orchards through careful investment of their labor. One Eastern newspaper reported that a Dutchess County, New York, man earned "astonishing profits from ten Apple Trees" in 1820. The ten grafted trees produced fifty barrels of Summer Russets, which brought $396 at market.[69] Stories like this one appeared with increasing frequency in Eastern journals in the 1820s, and in new Western agricultural magazines from the 1830s. The moral of these stories was clear: farmers who invested in good grafted stock and expended the necessary labor to raise healthy trees could yield handsome profits from even a small orchard.

But even among agricultural improvers, there were "scientific" defenders of the seedling tree as well. Some noted that popular grafted varieties appeared to decline in quality over the years. Noting the decline in quality over time of the once popular Bridgewater Pippin tree in Herefordshire, England, a decline that persisted even in young grafted trees, English horticulturist Thomas Andrew Knight concluded that tree varieties "aged" at a steady rate, and a young tree grafted from a hundred-year-old Bridgewater Pippin would suffer from the same deterioration in quality as its hundred-year-old parent plant. As a result, even the best grafted varieties eventually declined, so it was important to continue to develop new varieties from seed.[70] Belgian horticulturist Jean-Baptiste Van Mons shared Knight's belief that aging parent trees passed along their infirmities to their offspring, but not just when they were perpetuated by grafting. Trees sprung from the seeds of hundred-year-old trees also did. The solution, Van Mons, concluded, was to sow seeds only from young, vigorous trees and to do so over generations. He called this process "amelioration"—a favorite phrase of improvers like Ralph Waldo Emerson—and concluded that four generations of seedlings grown from young trees were required before superior new varieties emerged.[71]

While Knight's and Van Mon's belief that varieties of apples have a natural lifespan would not be considered sound science today, their theory was based on sound empirical observations and therefore gained some respectability among apple growers on both sides of the Atlantic. In Ohio, some improving farmers did notice that popular older varieties of grafted trees they imported from the East did not always thrive, or even that a variety that had been a reliable producer on their grandfather's farm appeared to be less reliable in their own time. The real factors contributing to this decline were external, not internal. A tree that produced impressive yields of high quality fruit in the climate and soils of the Connecticut Valley would not necessarily thrive in the Ohio Valley. Dramatic differences in results could also be seen for the same variety within the state, as a popular variety in Marietta might prove to be a mediocre performer in central Ohio. In addition, each new variety of apple contained its own set of vulnerabilities to specific viruses and pests, and those naturally increased over time as farmers planted more of them, creating environments for these pests to thrive.[72]

T. S. Humrickhouse of Coshocton, an early improving horticulturist in that county, sought to explain John Chapman's preference for planting trees from seed by suggesting that he was a devotee of Van Mon's theory. Humrickhouse had never met Chapman personally, as John had moved his operations a bit north and west by the time Humrickhouse had settled in Coshocton County.

But he had heard the stories of Johnny Appleseed from old timers in the region and seemed intent on offering a sympathetic portrait of the legendary apple tree planter. Rather than accept that John's aversion to grafting was the result of laziness or superstition, he sought in his account to invest in them a more scientific purpose.[73]

Perhaps most significantly, poor farmers, many of whom continued to follow a mostly self-provisioning lifestyle even after roads and canals had penetrated the interior, continued to see the value of an old, neglected, low-maintenance seedling orchard for meeting their personal needs. When the Ohio and Erie Canal was completed to Coshocton in 1830, one of the first canal boats to make it to the village contained a cargo of grafted apple trees. The cargo's owner hoped to find eager customers for these improved trees but was disappointed. Despite the new economic opportunities to sell goods to distant markets provided by the canal, the shift in behavior of Coshocton's farmers did not occur overnight. Most farmers in the area appeared content to keep their seedling orchards, and the disappointed peddler was forced to sell the whole lot off at a significant loss to the only interested party he could find.[74]

John Chapman continued to find customers for his seedling trees throughout the 1820s, but his business was not unaffected by forces rapidly transforming central Ohio. Agricultural societies and clubs were pressing improvement everywhere, and with better roads and the new canal, farmers had easier access to grafted stock than ever before. And, as noted above, a dramatic increase in population in the region was making it more difficult for Chapman to protect his squatter nurseries. In 1810, recently organized Knox County, counted just 2,149 white residents. Ten years later the population had increased nearly fourfold to 8,326. By 1830 the population had doubled again to 17,085. Richland County, which had become the base of Chapman's operations shortly after it was established in 1813, harbored 9,169 white residents by 1820, almost all of whom did not arrive until after the war. By 1830, Richland counted 24,006 residents.

GREAT BLACK SWAMP

John had managed to stay ahead of the crush of settlement during his first twenty years in the West. But by the second half of the 1820s, central Ohio had become settled enough that he needed to adapt or move. He chose to do a little of both. While he remained a regular presence in the Mansfield area into the early years of the next decade, he began to extend his activities westward.

Three lease agreements he entered in 1828 suggest that he was finding new opportunities in the northwestern section of the state, specifically the counties that were inundated—all, or in part—by the Great Black Swamp, the remnant of a glacially carved lake that sprawled across much of the Maumee River watershed between modern-day Toledo, Ohio, and Fort Wayne, Indiana. Despite its forbidding name, the Great Black Swamp was in reality a series of low-lying marshes and swamps, periodically broken by parcels of higher, drier ground. It was enough of an obstacle, however, to place a brake on westward settlement in northwestern Ohio for several decades. Crossing the swamp with a wagon was virtually impossible before the construction of a corduroy road from the edge of the Western Reserve to Perrysburg on the Maumee in 1825, and even then, during wet seasons, the journey could take weeks. In 1828, John Chapman entered into lease agreements with landowners in three counties on the southern edge of the Great Black Swamp. Mercer, Allen, and Van Wert were not formally organized as counties in 1830, but the white population that year of all three combined was just 1,737. Until the swamps were finally drained, growth in the region was modest, but along the southern border of the Great Black Swamp the population increased more than eight times—to 14,165 white persons—by 1840. For settlers even on good lands just south of the swamp, access to wider markets was quite difficult before the completion of the Miami and Erie Canal in 1845. It is no surprise then, that John Chapman began planting nurseries in this region, as it was a region filled with his kind of people—those who by choice or by limited means settled on some of the most isolated lands available.

The three leases Chapman entered into in that year appear to have been made at the time of the planting.[75] The first two were half-acre leases for a term of forty years, with the promise to pay the landowners forty apple trees within five years time. The last, on better lands further south from the swamp, was for "a certain piece of ground" of unstated dimensions, "lying below the Little Branch, below Shanesville . . . for the purpose of sowing apple-seeds on, and is to be cultivated in a nursery for the space of ten years, more or less." Property owner William B. Hedges struck a hard bargain, demanding one thousand trees from Chapman, "to be taken as they are suitable for the market or transplanting on equal proportion for the space of ten years . . . on average of One Hundred Apple Trees per year." Hedges further stipulated that if Chapman wanted access to any trees growing on the property after ten years, he would need to pay an annual rent of one hundred apple trees.[76]

In the same year he began entering leasing agreements in the counties below the Black Swamp, John unburdened himself of his remaining town lot in

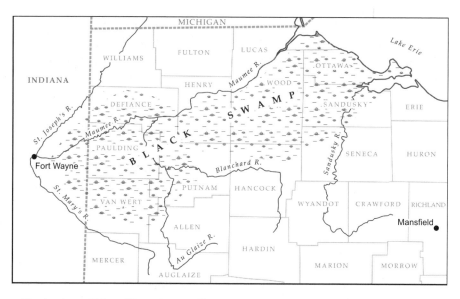

Northwestern Ohio and Fort Wayne, Indiana, ca. 1830. Claudia Walters, University of Michigan–Dearborn.

Mt. Vernon, selling it to Jesse B. Thomas for "Thirty dollars lawful money to me in hand"—just five dollars more than he had paid for it twenty years earlier when the "town" of Mt. Vernon was but a fiction. (The other lot appears to have been erased by a shift in the flow of Owl Creek.) Now a waterfront lot in a booming village, John's ability to make just five dollars on a twenty-year-old landholding again suggests he did not possess the stereotypical Yankee acumen for business.[77]

As Chapman began to shift his operations west, another group of central Ohioans were preparing for a longer, involuntary migration. The remaining Ohio Delaware, confined to a nine-mile-square reservation at Upper Sandusky relented to government pressure and the promises of land and annuities and agreed to relocate to Kansas. By this point the Ohio Delaware were utterly impoverished. A mysterious illness had swept through their reservation, killing the majority of their horses. Many members of the tribe had accumulated substantial debt with white trading houses, and when news of the imminent departure of the Delaware spread, merchants were eager to collect these debts. John Chapman may have played some small role in this last sad chapter of the Ohio Delaware. A few crumbling notes discovered in the papers of Mansfield merchant E. B. Sturges suggest that he employed a John Chapman to travel to the

Delaware reservation and collect debts owed by Solomon Johnacake and Billy Dowdee, two Delaware who had lived at Greentown on the Black Fork before the War of 1812.[78] Sturges was related by marriage to Chapman's Cooley family, increasing the likelihood that the John Chapman mentioned in these notes was indeed the apple tree planter.[79] He may have selected John for this delicate task because of his reputation for being on good terms with Ohio's remaining Native Americans or because John was constantly on the move, and he knew he would be passing through.[80] In either case, the departure of the Delaware in 1828 appears to have coincided with John's own gradual, voluntary migration away from a central Ohio region that was increasingly crowded and increasingly tied to national markets. When John sold off his 14.5–acre parcel in Wayne County in 1832 for just fifty dollars, ten less than he had paid for it a decade earlier, he had divested himself of all but one piece of central Ohio land.[81]

Despite these unprofitable land deals, John Chapman appeared to be getting by. He turned sixty years old in 1834 and was selling enough apple trees to accumulate a stash of money. New opportunities to acquire cheap land were opening up in western Ohio and northeastern Indiana, and Chapman was now in a position to pay cash for them. In April and May of that year he acquired two tracts of Indiana "canal lands" sold off to fund a state canal project. The first was a forty-two-acre tract north of the Maumee River purchased for $2.50 an acre, the second a ninety-nine-acre tract acquired for just $1.50 an acre. Also in 1834 he apparently acquired three "town lots" in the Hancock County, Ohio, village of Mt. Blanchard, on the upper reaches of the Blanchard River, in the region just below the Great Black Swamp. The "village" of Mount Blanchard had been platted but not settled, and the lots were undoubtedly for situating a nursery.

The middle years of the 1830s were an era of prosperity in the Midwest, and land prices in already settled regions rapidly escalated, especially in the years 1834–36. The cheap "canal lands" being offered up to fund the Wabash and Erie canal were an exception, although much of this land was swamp, and would not make for exceptional farmland until the Great Black Swamp was drained a few decades later. In March 1836, John obtained two more Indiana canal land properties in Indiana for the low price of $1.25 an acre, 18.7 acres in Maumee township, Allen County, near Fort Wayne, and 74 acres in Wabash township, Jay County, about fifty miles further south. What medium John used to pay for these lands is unknown, but he paid in full. Five months later, in response to growing economic uncertainty, President Andrew Jackson issued the Specie Circular, requiring that all public land be paid for at time of purchase in gold or silver.

Less than twenty years after the financial panic of 1818–19, another period of unprecedented economic growth quickly turned to a bust. The causes of this economic collapse were more difficult to identify than the last crisis, and today economists understand that the economic trouble was brought on at least in part by events in other parts of the world. But in the United States, Jackson's Specie Circular decision appears to have made an already tenuous economic situation worse, causing a run on specie from the banks, and encouraging specie-starved banks to call in their debts. The Panic of 1837, and a secondary panic in 1839, brought about five years of deflation and economic uncertainty. The downturn hit urban workers most, but even in the countryside farmers with too much debt felt its effects. Both ordinary Americans and their politicians, unable to see the invisible global market forces that were now exerting themselves on a nation that no longer operated in isolation, quickly began looking for villains. In politics, Whigs quickly pointed the finger at Andrew Jackson's war against the Bank of the United States, and Democrats, who had long blamed the evil national bank for the economic crisis of 1819, understood the panic as the last destructive act of that powerful institution in its death throes. Some saw national and personal moral failures at the root of the crisis. Transcendentalist minister Orestes Brownson denounced the "Spirit of Gain" as "the mother of all harlots and abominations of the earth," while others looked at the personal downfalls of some individuals and attributed it to their taste for "high living" and their efforts to "acquire wealth without labor."[82] For simple farmers who had always been ambivalent about markets, the seemingly inexplicable economic crisis was a reminder of the dangers and insecurities of the new economy. John Chapman, who had become the butt of many jokes about his austere lifestyle, weathered the economic downturn with no apparent ill effects. In the ensuing hard years, jokes about Chapman's extreme frugality were almost certainly retold with at least a speck of admiration for his virtue.

Back in Massachusetts, another man who had professed a commitment to living simply but had never quite lived it found his fortunes reversed by the economic crisis. Bronson Alcott had by the middle of the 1830s gained some fame for his ideas about education and his commitment to vegetarianism, but that fame did not save him from economic troubles. He had accumulated debt fitting out his school with a substantial library and busts of famous philosophers, and when hard times and a few missteps resulted in declining student enrollments, Alcott was forced to declare bankruptcy and sell off most of his possessions, including his beloved library.[83] The crisis appeared to steel his determination truly to live the life of simplicity he had long espoused.

He began plans for Fruitlands, a new utopian community where money would not exist and the individual would not be compromised by any association with exploitation. The residents of Fruitlands dressed in simple linen tunics produced without the exploitation of slave or animal labor (cotton, leather, and wool were all banned) and subsisted primarily on a diet of apples and bread. The experiment, on an eighty-acre farm near Harvard, Massachusetts, set very high standards for itself but did not last a year. Life at Fruitlands was largely defined by self-denial. Coffee, tea, sugar, alcohol—all products of the marketplace—were banned. All animal products, not just the meat of slaughtered livestock but also the eggs, milk, and wool taken from animals without their consent, were off limits. "Pluck your body from the orchard, do not snatch it from the shamble," Alcott declared. Even the labor of the ox could not be exploited for plowing the fields, although the colonists temporarily abandoned this principle in the name of expedience. Fruitlands never housed more than sixteen residents, and most of the time Alcott, his wife, and four daughters made up about half of the entire colony. Toward the end of the commune's short life Alcott came to the conclusion that his marriage and family were also obstacles standing in the way of his pursuit of perfection, and he momentarily considered abandoning them.[84]

On a cold late fall day in 1843 as Alcott walked across the fields of Fruitlands, a coarse linen tunic draped over a gaunt frame sustained on a diet of apples and bread, contemplating a life of celibacy, he must have looked, momentarily at least, eerily like John Chapman, then traversing the Indiana-Ohio border, some 850 miles to the west. That Alcott and Chapman were even aware of each other's existence is unlikely. But it seems obvious that the lives of both were deeply influenced by the same forces—the perfectionist fires that first sprung from religious awakenings and deeply ambivalent feelings about the market revolution and its transforming effects on society.

John Chapman made no new land purchases in 1837, but he secured the patents (presumably by having them surveyed) on the two he had acquired a year earlier, and the only land he sold in the ensuing depression years was the three Mt. Blanchard town lots, in 1839.[85] Furthermore, he continued to make interest payments on his last remaining Ohio school lands lot near Mansfield. In 1838, he acquired another forty-acre Indiana canal lands tract near Fort Wayne.[86]

By this time, Chapman appears to have made the Fort Wayne, Indiana, area his base of operations, and his visits to Mansfield were increasingly rare. At some point—perhaps in the wake of the panic—he invited his sister Percis and her poor family to come out and reside on one of his Indiana properties. Rosella

Rice described the Brooms in condescending terms as a helpless but upbeat family—Percis, "a spry little woman who always called me 'honey'"; William, "an easy old gentleman"; and their children, "pretty little girls with their sunny hair and their laughing brown eyes." They set out for Indiana in midwinter but made it only as far as the Rice home when their possessions, haphazardly tied on "a poor caricature of a beast" began spilling off. Rosella's father, "to get rid of them," built them a primitive sled that the horse could drag, gave them some provisions and a five-dollar note, and "sent the whole kit on their way rejoicing." When John next visited the Rice family, "he very promptly laid a five-dollar bill on my father's knee, and shook his head very decidedly when it was handed back; neither could he be prevailed upon to take it again." John clearly felt it was his responsibility to look after his younger sister and her family.[87] The land John Chapman allowed William and Percis and their children to move onto was unbroken, and it would be up to William Broom to clear it and build a home there. Several years later he made a claim against John Chapman's estate for the improvements he made to the property.[88]

By all accounts John continued to live the meanest of existences. Those who encountered him in northeastern Indiana described him in much the same way as those who knew him in Ohio. His clothing consisted of cast-off rags, and there was often too little of it for the climate. Even on cold days, he could be seen striding across the countryside in bare feet or the most ragged of footwear, and in his shirtsleeves. Wherever he went his commitment to an ascetic lifestyle drew commentary.

Religion, too, continued to remain an important part of his life. He was seen stretched out on the ground under the shade of a tree at a Methodist camp meeting on the outskirts of Fort Wayne on summer evening, and according to one source he had a repeat encounter with the preacher he had confronted on the Mansfield town square a few years earlier. According to Fort Wayne resident John Dawson, the minister in question was Adam Payne, and far from being well-dressed and respectable, Payne seemed to be an imitator of the infamous wild-haired, unkempt preacher Lorenzo Dow. If Chapman and Adam Payne did indeed have a second preach-off on a street corner in Fort Wayne, it was a contest between two men trying to outdo one another in their primitivism as a show of piety. But most who encountered John Chapman in this region did not perceive him as overly righteous and suggested that he rejected the hellfire and brimstone style of religion, telling one Paulding County family that "the worst part of hades would not be worse than smoky houses and scolding women."[89]

JOHNNY APPLESEED.

Page 27.

The earliest drawing of John Chapman, artist unknown, was sketched from a description given by Rosella Rice. While consistent with the most credible physical descriptions of John Chapman, the grafting knife in his right hand is the artist's own innovation. Chapman grew trees from seed and did not practice grafting. Horace S. Knapp, *A History of the Pioneer and Modern Times of Ashland County* (Philadelphia, 1863), frontispiece.

LOG CABINS AND HARD CIDER

The Panic of 1837 and ensuing depression did not convince the mass of Americans to abandon markets and money and return to a simple self-provisioning lifestyle. Americans continued to enjoy and celebrate many of the technological advances and the increased material comfort that the transportation and market revolutions had brought. But it did rekindle a bit of nostalgia for "simpler times." This nostalgia was successfully exploited by the Whig Party during the election campaign of 1840. By that year, many leading Whig politicians had come to recognize that they could not win the highest office in the land without embracing the populism that Andrew Jackson and his party had so successfully harnessed since the mid-1820s. As a result, they settled on old war hero, William Henry Harrison, for their presidential candidate.

The Whig campaign stumbled upon an opportunity to exploit this nostalgia for simpler times almost by accident, when a journalist for a Democratic newspaper, suggesting that Harrison was too old to run for president wrote, "Give him a barrel of hard cider, and settle a pension of two thousand a year on him, and my word for it, he will sit the remainder of his days in his log cabin by the side of his 'sea coal' fire, and study moral philosophy."[90]

Harrison supporters seized on the log cabin and hard cider elements of the insult to suggest that Van Buren and his supporters were elitists who disdained the lifestyle of the simple self-provisioning farmer. They quickly rebranded Harrison the "log cabin and hard cider" candidate, a perverse political twist on many levels. Far from being born in a log cabin, Harrison was born on a James River plantation, a descendant of one of Virginia's elite, slaveholding families. Even most of his days as a military officer in the West were spent at Grouseland, a magnificent estate, surrounded by grafted fruit trees and gardens, which he had built in Vincennes during his time as governor of the Indiana territory.[91] In a more transparently ironical way, Harrison supporters built log cabins on the decks of steamboats and powered from town to town to win votes for their log cabin hero.[92]

Temperance reformers, who were a staple part of the Whig coalition, were appalled by the gimmick and threatened to abandon the party for promoting alcohol consumption. "*Intemperance has become the badge of a political party!*" Harrumphed the *New York Evangelist*. "Yes, intelligent men—men who have enjoyed the benefits of Christian teachings—and who live in a land of gospel light—are called upon to exhibit their enthusiasm for political strife, by drinking *hard cider*, made harder by hard *brandy*, for the Glory of General Harrison!" The *Evangelist* predicted that through this craven political gimmick, "More than ten thousand men will be made drunkards in one year by this hard cider enthusiasm."[93] A writer in another New York paper, declaring it "a burning shame that the flag of my country waves over such mockery and abomination, as though her stars and stripes were not insulted by being associated with such iniquity," issued a warning to the Harrison campaign. Should these grog-dispensing log cabins be opened on "Sunday, either day or night," the Whigs would lose the votes of so many temperance men that it would negate the effect of this pandering.[94]

But however much the log cabin and hard cider campaign exasperated temperance Whigs, the strategy worked. Americans troubled by the economic malaise that had fallen upon the country in 1837 embraced a nostalgia for a simpler time when their fates were not tied to mysterious market forces beyond their control. Log cabins and hard cider were a perfect symbol of that lost past. The celebration of hard apple cider represented not simply a celebration of the disappearing self-provisioning lifestyle, but it was also a protest against do-gooder moral reformers bent on telling ordinary people how to live and what to drink. It did not seem to matter that the Whig Party's soft money, pro-development economic policy promised to accelerate the market revolution, or that those moral do-gooders were most commonly associated with the Whig party. This was political triangulation at its finest. Voters were won over by the celebration of the seedling apple orchard and its homegrown product. Despite the theatrical celebration of simpler times, it would be folly to suggest that Harrison's campaign succeeded by completely deceiving voters as to its real agenda. Most voters certainly understood hard cider and log cabin celebrations as one great play. Such nostalgic celebration of the past did not mean the celebrators had an authentic desire to turn back the clock. Whigs could afford to give a wink and a nod to the simple pleasures of earlier times while pushing the nation into a new age.

Perhaps there was no greater example of the shallowness of hard cider enthusiasm than the fact that during these same depression years the movement to criminalize apple-pilfering began gaining steam. As late as 1819, while traveling in eastern Pennsylvania, Englishman William Cobbett confidently asserted that taking another man's apples would never be considered a crime in the United States. If this were true in regions already attached to urban markets, it was even more true in Ohio. Referring to his boyhood in eastern Ohio in the late 1810s and early 1820s, William Cooper Howells commented that there "were plenty of apples in the orchards . . . where they were always free to the passerby."[95] Howells recalled that even a one mile trip to the flour mill could "spoil a day's work" because along the path he would have to pass two orchards, a good fishing creek, and a good swimming hole. Howells was inclined to ask permission of the owner before taking fruit, which was always granted, but many other travelers, both children and adult, never bothered with such niceties. The reason was that once mature, orchards produced fruit in such abundance that the typical farm family short on labor and with limited access to markets could not possibly sell or process it all, and much would go to waste. On his own family farm in 1820, "the peach crop was too great for us to manage, and much of it went to waste" despite the valiant efforts of the family to gather and dry as much of the crop as they could.[96] As a result, people tended to see orchard fruit as a providential bounty free for the taking when God delivered it up.

To be sure, many saw taking fruit without asking to be a sin and bemoaned the prevailing attitude "that every body has a legal right to eat as much fruit as he wants, wherever he can find it." Essays on the moral instruction of children often used the example of taking apples from orchards without permission as a kind of gateway sin that could lead little boys down the wrong path. Noah Webster's *American Spelling Book* used variants of a fable about a boy who stole apples in many editions of the primer. But in most of the cautionary tales about the sin of stealing apples, when the boys learn to ask the owner for permission they are granted permission to take "just as many as you want." Even the moralists recognized that orchard fruit in season was in such great abundance that it would be stingy to deny anyone permission to fill their pockets, so long as they asked.[97]

As far as the law was concerned, taking apples from a roadside orchard was a trivial offense—at most an instance of trespass, for which the owner was only entitled to sue for the value of lost fruit, which in any given case would be so small as to not be worth the trouble. But as good roads and canals connected farmers to urban markets, and improvement-minded farmers invested more of their labor and resources into carefully cultivated, grafted fruit orchards with marketable winter apples, they began to perceive the passerby who pilfered a shirttail full of apples in the same light they did the pickpocket. Market-minded farmers grew increasingly frustrated that the law did not agree. As early as 1832, a court case in New York gave horticulturalists some hope that the legal system and the public might begin to take their grievance seriously. The case involved an apple-pilferer who took a farmer to court for assault. The pilferer had been caught in the act by the orchard owner who was holding a horsewhip when he demanded that the thief put down the fruit. When the brazen scoundrel refused, the orchard owner took the whip to him. The plaintiff's lawyer confessed to the jury that he himself had on many occasions taken fruit from other men's orchards, that no doubt the majority of the jury members had done so as well, and therefore the assault with horsewhip was entirely unwarranted. The jury disagreed and found for the farmer defendant. The story circulated in agricultural journals, who read into the jury's decision "the pleasing hope that we were on the eve of a revolution in regard to the plundering of fruit, and that a great improvement in public sentiment is taking place on this subject."[98]

The revolution did not occur overnight, but as farmers began organizing state agricultural societies, they began to lobby state legislatures to pass laws that treated apple-pilfering as a crime beyond mere trespass and imposed penalties far beyond the value of the product stolen. Eastern states led the way. The movement to criminalize apple-pilfering was at base a movement to press Americans to begin to recognize fruit-raising as a legitimate industry and fruit as a valuable form of property, worthy of the same protections extended to livestock and manufactured goods. It was also an effort to exterminate the old self-provisioning culture that John Chapman's seedling apple trees represented.

It is no surprise, then, that in his last years, John continue to press into regions more isolated from these markets and their values to find new customers. But market penetration was moving at a faster pace than his migration. At least one report suggests he traveled farther west to plant apple trees, into the state of Illinois in the early 1840s. But age was also catching up with him—John turned sixty-six years old during the fall of Harrison's hard cider campaign—and if he did make such an exploratory trip, there is no evidence that he ever returned to take care of seedling nurseries he may have planted there. To be sure, even in areas where market penetration was significant and grafted fruit stock was available to new settlers, his seedling apple trees could still find some market with poor farmers who hoped to use them to get a start on an orchard. But John continued to operate outside the values of the marketplace. He sustained himself on the profits of his apple trees and appeared to be more concerned with being useful than being successful, in the terms of the new market economy. His willingness to set his price based on the needs and resources of his customers—even giving them to the poorest for free—was at odds with the notion of a fixed market price for goods. Near the end of his life there was little ambiguity about John Chapman's answer to the question of his age—how a man might serve God instead of Mammon. Few of his neighbors answered the question in his way, and that was the source of his fame.

CHAPTER SIX

Yankee Saint and the Red Delicious

In the fall of 1842, John Chapman turned sixty-eight years old. He was by this point a fixture in Fort Wayne, Indiana, and in the counties bordering the Great Black Swamp to the east. And he was still busy planting and selling apple trees. No doubt he had long ago reconciled himself to many aspects of the new economy that the market revolution had brought. He was quite accustomed to cash transactions, and he had acquired a substantial amount of property as a way of protecting his nurseries, perhaps also as a place to sink the surplus cash he had been accumulating but disliked carrying around. When planting on land he did not own, he would enter formal leasing arrangements and record them at the local county courthouse. But he never reconciled himself to the new materialism, and he continued live, eat, and dress with extreme frugality. As age overtook his body he was more inclined to board indoors rather than sleep rough.[1] And while he might pay for a meal in a tavern, he was, according to one who remembered him in these years, "exceedingly penurious" and complained that six pence was more than sufficient price for a meal.[2] His charity toward the poor, however, was unabated. Another who knew him during these years noted that he was "known to give for benevolent purposes 5–10–15 and even $50. He generally had money."[3]

In the fall of 1842, Chapman made one final trip down to the Marietta, Ohio, region to visit his half-brothers and sisters. We don't know how he traveled, but it is probable he walked the whole way. Ohio had experienced revolutionary change since he first crossed into the territory about 1801, growing from roughly forty thousand residents to more than one and a half million. If he passed through his old haunts of Mansfield and Perrysville in Richland County along the way, he found a population that had increased about five times in just the last twenty years.[4] At some point on this journey John Chapman had to cross the National Road. By 1842 the road was sixty feet wide and fully macadamized across the state. Pedestrians walked along the margins while steady traffic of coaches, carriages, wagons, and livestock crowded the road.[5] At Zanesville where the National Road crosses the Muskingum River, John had in 1801 encountered a tavern and a few log homes. Two score years later, John entered a small industrializing city of about eight thousand, many employed in iron foundries, salt works, and potteries. More than one in three adult males in the county was employed in nonfarm work by 1840.[6]

John also encountered a very different river. By 1842, the Muskingum had been tamed by a series of locks and dams that regulated water flow from Marietta on the Ohio River to the canal town of Coshocton thirty miles north of Zanesville. Goods could now be transported from Lake Erie to the Ohio River by an all-water route, and passengers could travel upstream under steam power all the way from Marietta to Coshocton. At Zanesville a half-mile-long stone-sided canal carried steamboats around the old falls and stepped them up and down through a set of double locks.[7] John had made the journey from Zanesville to Marietta and back many times over the years. In the early days, the return journey required following narrow paths along the steep ridges above the river or paddling and poling a canoe against the current, negotiating a series of small rapids and falls along the way. By 1842, he had new choices. He could make the journey by steamboat, but the ever frugal John probably declined and instead walked the sixty miles along a newly completed road that followed the course of the river.

In Washington County, John stayed with his half-brother Nathaniel, the boy he had taken with him into the wilderness of northwestern Pennsylvania forty-five years earlier. Now married and with a large family of his own, Nathaniel owned a mill along Olive Creek and had acquired several parcels of farmland in Washington and Morgan County.[8] His baby sister Sally also lived in the vicinity with her husband and family. Jonathan Cooley Chapman—according to family tradition a deaf mute—lived alone near Broken Tree, a few miles west of the Duck Creek lands his father had first landed on. His brothers Parley and Davis also had families and were living in the area.[9]

W. M. Glines, a neighbor of Sally's recalled John's final visit in a memoir. According to Glines, Sally's husband John Whitney told the apple tree planter that during a recent storm lightning had struck a massive black oak tree on the far side of his farm, splitting the tree into rail-sized pieces, which Whitney had then sent his sons to collect and make into a much needed rail fence. Chapman was so taken with the story that he demanded to see the place where this had occurred. Glines, Chapman, his brother Nathaniel, and Whitney set off through fields and briars to the location, John peeling off his shoes and going the rest of the way barefoot when they came to a stream.

When they came to the charred remains of the tree, John Chapman launched into "a sermon on the wonderful Providence of God to man," suggesting that God knew the Whitney family was short on labor and desperately needed a rail fence and that the lightning strike was God's doing. Whitney "hung his head for a moment, then replied, that he always tried to feel thankful to God for his kind care over him and his family, but that he had never heard of His making rails for anybody before." When Chapman insisted that he "must receive all such Providences as special favors from God," Whitney deflected and, recalling the story that John had told him probably a dozen times before about getting stuck in a snowstorm in the middle of the woods and having to shelter in a snow bank, asked him why, if God intervened in people's lives in such direct ways, did He not save him from that miserable situation? John responded that he had been "a great fool for putting myself in that situation" and that God in his mercy had "sen[t] snow enough so that I could dig a hole in it and secure myself from freezing."[10] Whitney was no doubt a pious churchgoing Protestant and was not inclined to press his challenge. But John's insistence that God intervened daily in the lives of people was met with a touch more skepticism in the modernizing world of 1842 than it might have been forty years earlier.

Before bidding goodbye to his Washington County family, a niece who understood her uncle's values quite well presented Chapman with a gift—a shirt she had made just for him from some old scraps of cloth. The shirt was one half muslin and one half calico. The muslin, coming from an old feed sack, "had two large letters, perhaps A.D" printed across its front. Chapman was reportedly delighted with the gift, no doubt because it was handmade and a testament to the virtue of frugality. He wore it proudly, and we can imagine that such attire drew more stares in this age of stagecoaches and steamboats than it would have in earlier days.[11]

His Washington County family never saw him again. He spent his last few years tending his nurseries and engaging people on the subject of politics or the afterlife whenever he could find a patient ear. In March 1845, John Chapman passed away. The *Fort Wayne Sentinel* noted that "his death was quite sudden. He was seen on our streets a day or two previous."[12] But other accounts tell of him falling ill and finding shelter at the home of William Worth, who cared for him in his last days. According to a witness, he was wearing at the time of his death "a coarse coffee-sack, with a hole cut through the centre through which he passed his head. He had on the waists of four pairs of pants. These were cut off at the forks, ripped up at the sides and fronts thrown away, saving the waistband attached to the hinder part. These hinder parts were buttoned around him, lapping like shingles so as to cover the whole lower part of his body, and over all these were drawn a pair of what was once pantaloons."[13] This erratic collection of scraps was not enough to protect him from the cold winds that whip across the plains of northeast Indiana. His death was attributed to "the winter plague."[14]

Chapman's death warranted more ink in the local Fort Wayne newspaper than that of an immigrant laborer who died the same day. "Dies—In this city on Tuesday last, Mr. Thomas McJanet, a stone-cutter, age 34 years, a native of Ayrshire, Scotland" was the full obituary for Mr. McJanet. Chapman's notoriety made him worthy of several paragraphs. The *Fort Wayne Sentinel* reported that John Chapman "was well known through this region by his eccentricity and the strange garb he usually wore. He followed the occupation of a nurseryman and has been a regular visitor here upwards of twenty years." The obituary also indicated that he was a follower of Swedenborg and "he is supposed to have considerable property, yet denied himself the most common necessities of life." The paper credited his religious beliefs for this contradiction. As to other details of his life, the *Sentinel* could only repeat local speculation and rumor. He might have been born in Pennsylvania; he might have had family near Cleveland; and most interestingly, "he was not less than eighty years old at the time of his death—though no person would have judged from his appearance that he was sixty." Chapman was in fact seventy-one. One is left to wonder what aspects of John Chapman's person made him seem older than he actually was and which made him seem younger. Perhaps his gaunt frame, unkempt hair and beard, ragged clothes, and sun-dried skin were responsible for the editor's adding a decade to his age, while his physical fitness made him seem younger. Or perhaps it was his disinterest in money and material things and his quaint ideas, about God making rail fences for families in need, that led the editor to conclude he was at least an octogenarian. No one of later generations could possibly be so out of step with the times.[15]

Yet while Chapman's local notoriety was worthy of several paragraphs in the Fort Wayne newspaper, news of his death did not spread quickly. Neither Henry Howe nor T. S. Humrickhouse was aware of what had become of John Chapman when they published stories about him two years later. And eight years later, when Lansing Wetmore of Warren, Pennsylvania told the story of John Chapman's winter crossing of the Alleghenies, he did not even know that John had earned the nickname Johnny Appleseed in Ohio in subsequent years. Instead he confused him with another John Chapman who made a small splash in national politics in the 1840 campaign.[16] It was many years before the collection of locally told stories about John Chapman began to coalesce into a coherent story of his life; basic facts about such things as his birthplace and date, and even the date and exact place of his death, remained elusive or contested for a century after his passing.

It took more than a decade to settle John's estate. The estate records provide a clearer understanding of John Chapman's financial situation at the time of his death. In northeastern Indiana he held title to four parcels of land, totaling about 175 acres. Three of these were fully paid for, but he still owed $120 in payments and taxes on one forty-two-acre parcel, and the land was valued at about that much.[17] He still held legal title to one of his school lands leases near Mansfield and perhaps one or two other small Ohio properties. And he had property in apple trees. One nursery of two thousand apple trees was assessed at a value of forty dollars, or two cents per tree, and another of fifteen thousand trees, assessed at three cents a tree, deemed to be worth $450. In Mercer County, Ohio, an estate administrator was able to sell about 440 of Chapman's seedling trees for six cents apiece. Yet the true value of seedling trees was fleeting. A neglected seedling nursery could soon become choked with weeds and was vulnerable to fire and flood; even saved those perils, it had a short life before the trees became too large for transplanting. As it took years to settle Chapman's estate, the value of any remaining seedlings evaporated. John Chapman also owned at the time of his death "one gray mare," valued at $17.50. Furthermore, Chapman was owed money by several people, presumably unpaid bills for trees from his nursery.

Against these assets were quite a few claims. Chapman's brother-in-law William Broom claimed $127.68 for improvements he had made on one of John's canal land parcels. This included clearing and fencing four acres of land, "building a Log House 18 by 21 feet," and "scoring and hewing timber" for a barn. Joseph Hill claimed $104 against the estate for periodically providing John with board between 1837 and 1844: fifty-three weeks total, at a rate that was sometimes $2, sometimes $1.50 per week. Richard Worth claimed $7.50 for boarding John for five weeks, spread across five years, and also for the funeral expenses he incurred for laying him out. And then there were the curious bills of exchange, each for $100, signed back in Franklin, Pennsylvania, in 1804, one to the children of his sister Elizabeth, the other to a Nathaniel Chapman. The notes were filed against the estate in November, eight months after his death, presumably by some member of his family. His sister Percis was most likely the holder, as she was the only relative living nearby. The court recorded the notes but never accepted them, so they were not paid. John had done his best to make them meet legal requirements for enforcement, having two witnesses sign the notes. But they may have been rejected because they lacked specific information about interest payments and when they were to come due.[18] When the final settlement was sealed in January 1856, almost eleven years after John's death, claims against the estate exceeded its total value, and most, but not all claims were paid. John Chapman did not die a wealthy man, but he was not impoverished either. From a dollars and cents perspective, John Chapman "got by," and that seemed to be exactly what he intended to do.

He picked a very fitting time to pass into the next world. The depression that had followed in the wakes of the panics of 1837 and 1839 had begun to lift by 1843, and two years later Midwestern farmers seemed more enthusiastic than ever about advancing from self-provisioning lifestyles to commercial agriculture. In January 1845, a statewide journal devoted to the advancement of agriculture in Ohio, the *Ohio Cultivator*, published its first issue, indicating that a critical mass of improvement-minded farmers had established themselves in the state. Also in that year Andrew Jackson Downing published the first edition of his catalog of American orchard fruit, *Fruits and Fruit Trees of America*. And there were other signs of the changing times. Just one week after John Chapman's passing, the Ohio state legislature finally passed a law establishing severe penalties for damaging fruit trees. The law first applied only to Cuyahoga County but quickly extended to the other Western Reserve counties and a handful of other counties with substantial Yankee populations. One year later, it was extended to the entire state, after its champions won over the skeptics by reducing the maximum imprisonment penalties. By the stipulations of the modified 1846 law, an apple pilferer faced fines, but not imprisonment, if the theft or damage was less than fifty cents in value and under no circumstances faced a penalty greater than sixty days in jail.[19] The political leadership of Ohio had endorsed the idea that apples were not simply God's providential bounty but a commodity with a measurable market value. By the end of 1846, nineteen Ohio counties had formed agricultural societies to protect the interests of Ohio's commercial farmers, and in the following years new ones sprung up annually. Virtually every county in Ohio had an agricultural society before the Civil War.[20]

YANKEE SAINT

At the time of Chapman's death, ungathered stories about the wandering apple tree planter still circulated across western Pennsylvania, Ohio, and Indiana.

The first to be captured in print in the United States (recall that Chapman made it into a New Church annual report in Britain in 1817) appeared in the *Ohio Cultivator* when T. S. Humrickhouse, the Coshocton County horticulturist, wrote up the local stories he had gathered from old timers about John Chapman and published them in a letter in 1846.[21] One year later, another account appeared in Henry Howe's *Historical Collections of Ohio*.[22] Between 1847 and 1861, short pieces about the eccentric apple tree planter, one misidentifying him as "Sammy Appleseed," appeared in agricultural and religious periodicals sporadically across the North.[23] More accounts appeared in the 1860s, when the first Ohio county histories were published.

Examined collectively, these early locally originated stories told by those who claimed to know him (or to have spoken to someone who did) had common elements but did not point clearly to a single interpretation of his life. Many focused on his meekness, his monomania for apple tree planting, and his benevolence, but others contained elements that were common to frontier folk traditions in the trans-Appalachian West. These included stories about his extraordinary skill at chopping trees or his ability to endure pain and cold. Some spoke of his religious conviction in admiring tones, while others poked fun of his peculiar and unconventional beliefs. His habits of dress and his extreme reluctance to harm any living thing, even a mosquito, were the subjects of derisive stories. Some called him lazy, others suggested he was crazy, and Catherine Stadden described him as impractical. He was commonly recognized as of New England origin but only imperfectly fit the archetype of the Yankee peddler. He possessed Yankee thrift, and sometimes Yankee ingenuity, but lacked the shrewdness and ambition typically associated with the type. John Chapman had moved easily among Ohioans of all stripes, including those who came from the upland South and the Scots-Irish and Germans who poured in from the mid-Atlantic states. He was also known by all classes of frontier dweller and counted landed town boosters, family farmers, and simple backwoods squatters among his friends. In the first fifteen years after his death, "Johnny Appleseed" was nothing more than a collection of tall tales about a peculiar frontier character.

But by 1880, John Chapman had found his hagiographers, and a more coherent narrative about the meaning of his life emerged. The narrative was primarily sentimental and reverential, but it retained some elements of the comic. In these newly published stories and poems, John Chapman emerged as a Yankee saint. Where some had seen unorthodox religious belief, the hagiographers saw extraordinary piety; where some had seen an irrational monomania for planting seedling apple trees, they saw intentionality of purpose; where some had mocked his seeming indifference to money and collecting debts, they saw charity; and where many had ridiculed him for his peculiar, ragged dress and strange eating habits, they saw frugality and thrift. For his new champions, Johnny Appleseed was a propagator not just of apple trees but of northern middle-class reformer values, specifically, piety, usefulness, benevolence, and frugality. Three individuals were especially important in the invention of this Yankee saint.

Rosella Rice, a lifelong resident of Perrysville, Ohio, had personal memories of John Chapman's visits to her home when she was a little girl. In adulthood, Rice wrote down all of her memories of him and gathered stories from other locals. By the 1860s, Rice had published a sentimental tragedy called *Mabel, or the Heart Histories* and began regularly publishing fiction in popular national women's magazines like *Godey's Lady's Book* and *Arthur's*, sometimes under her own name and also under the pseudonyms Chatty Brooks and Pipsissaway Potts.[24] She provided material on Chapman for several of the early county histories published in central Ohio and a sketch of Chapman for another local writer, James M'Gaw, who published a fictionalized account of the story of the Seymour (aka Zimmer) family, who had been killed by Indians during the War of 1812. M'Gaw made Chapman a central character in his work of historical fiction, *Philip Seymour*, but he had never known the apple tree planter. He depended on Rice and other locals for stories about him. Rice also published a few accounts of Chapman in national magazines in the early 1880s. In recounting John Chapman, Rice's language was hagiographic, as is evident in this eulogy she wrote for him, which was included in an 1883 edition of *Philip Seymour*: "His bruised and bleeding feet now walk the gold-paved streets of the New Jerusalem, while we so brokenly and crudely narrate the sketch of his life—a life full of labor and pain and unselfishness; humble unto self-abnegation; his memory glowing in our hearts, while his deeds live anew every springtime in the fragrance of the apple-blossoms he loved so well."[25]

But it was an article published in *Harper's Monthly Magazine* in November 1871 that catapulted John Chapman to national fame. "Johnny Appleseed— A Pioneer Hero" was penned by W. D. Haley, a New England–born Unitarian minister. Haley shared the reformist values of other Yankees of his generation, which had run him into trouble when he expressed antislavery views a bit too forcefully from the pulpit of an Alton, Illinois, church in the years before the Civil War.[26] After the war, Haley, who had abandoned the pulpit for a career as a journalist, became a champion of the Patrons of Husbandry, also known as the Grange movement. By 1870 Haley was actively organizing local Granges across Ohio and the Midwest.[27]

The Grange movement was the farmer's response to a new crisis in agriculture that had emerged after the Civil War. By this time, American farmers had embraced the doctrine of agricultural improvement thoroughly, adopted new methods for growing crops, and invested in new technologies that allowed them to increase yields with no additional labor. American agriculture reached new heights of productivity, but prices for food crops declined as a result. In addition, new technologies meant new dependencies, and farmers fell under the thumb of monopoly-controlled railroads and grain-storage elevators for getting their crops to market. While the Patrons of Husbandry celebrated the virtue and importance of the farmer in American society, they came to conclude that in an unregulated national market, the farmer was the loser. "Faith, Hope, Charity, Fidelity" was the motto of the Patrons, which also encouraged its members to "buy less and produce more" by institutionalizing the old frontier practices of cooperation over competition.[28]

Haley's Grange activism and his sympathy for the new challenges faced by America's rural communities formed a backdrop for his depiction of John Chapman. He noted that "the railroads have destroyed the romance of frontier life, or have surrounded it with so many appliances of civilization that the pioneer character is rapidly becoming mythical." Haley had no interest in the "rapine and atrocity" of the frontier experience celebrated in so many "dime store novels" by this time. In recounting John Chapman's story he sought to celebrate "sublimer heroisms than those of human torture, and nobler victories than those of the tomahawk and scalping-knife." In Haley's mind, John Chapman deserved to be remembered as a saint of the farmer's frontier, as one of "the heroes of endurance that was voluntary, and action that was creative not sanguinary." To be sure, neither Haley nor his brothers in the Grange desired a return to those primitive times; they had thoroughly embraced improvement. But Haley nonetheless saw in John Chapman certain admirable values that persisted in the nation's rural farm communities. Despite his ragged appearance and primitive agricultural methods, John Chapman embodied the values of piety, frugality, and charity championed by the Grange.[29]

Haley spent a few years writing for a local newspaper in Mansfield, Ohio, where he gathered local Johnny Appleseed stories for his essay. Rosella Rice was clearly one of his sources, as some of the language he used was identical to things Rice wrote about Chapman before and after the *Harper's* article appeared. Rice may also have borrowed from Haley, as some of her post-1871 pieces on Chapman contain phrases that appeared first in Haley's account. The version of Johnny Appleseed who emerges in the writings of Rice and Haley was a gentle old man who lived his life for others.³⁰ He was frugal to extremes. His generosity and compassion were unmatched. "He was never known to hurt any animal or to give any living thing pain—not even a snake. The Indians all liked him and treated him very kindly. They regarded him, from his habits, as a man above his fellows."³¹

Haley and Rice were also fundamental in turning John Chapman into a kind of magical Santa Claus responsible for almost all the apple trees planted across Ohio. Rosella Rice claimed that virtually every "orchard in the white settlements came from nurseries of Johnny's planting." Rice and Haley also painted highly colored visions of John Chapman's shabby seedling nurseries, describing them not as patches of hastily sowed and quickly neglected seedlings but as shrines of natural beauty. "The sites chosen by him [were] such as an artist or a poet would select," Haley wrote, "rich, secluded spots, hemmed in by giant trees, picturesque now, but fifty years ago, with their wild surroundings and the primal silence, they must have been tenfold moreso."³² Haley's description is based on Rosella Rice's claims that she knew where many of these nurseries were and that Chapman planted his trees in natural temples, framed by the sweeping arches of giant sycamores—hardly a practical location for raising sun-hungry seedling trees.

When Haley's piece appeared in the nation's most widely read magazine, it prompted a handful of rebuttals and responses from people who knew Chapman. Most focused on minor errors in fact or description, with none as harsh as Catherine Stadden's dissent, which was printed after the *Harper's* article appeared (see chapter 3). But these rebuttals, published in local newspapers and in the reports of local historical societies, stood no chance of upending the myth creation.

Perhaps no early celebrant of Johnny Appleseed was as effective at promoting the northern sentimentalist reformer's Appleseed narrative as Lydia Maria Child, who published "Apple-Seed John," a children's poem about him, in *St. Nicholas* magazine in 1881. Born in Medford, Massachusetts, in 1802, Child is most remembered today for her iconic poem about Thanksgiving, "Over the River and through the Woods," which she penned in 1844. In her own time, she became very well known for her advice manual for women, *The American Frugal Housewife*, first published in 1829. Subtitled *"For those who are not ashamed of economy,"* the book offered practical tips to housewives needing to cut their expenses. In her obsession for keeping and finding uses for every candle stub and even the smallest scraps of fabric and twine, she was a kindred soul to Chapman. For Child, as for Chapman, frugality was religion and had the power to improve the world. "True Economy," Child declared, "is a careful treasurer in the service of benevolence; and where they are united respectability, prosperity, and peace will follow."[33]

The *Frugal Housewife* found an enormous audience and was in its thirty-third edition by 1855. But Child did not reserve her reformist enthusiasm to the household. Politically, she embraced a host of radical causes, from abolition and women's rights, to anti-imperialism and the rights of Native Americans.[34] Yet she never abandoned her writings for children and the domestic sphere, and in her poem "Apple-Seed John" she saw an opportunity to connect her passion for simplicity with living a life of intentional benevolence. In the poem, Child portrayed Chapman as a man with a clear vision of how he would make the world a better place, one not recognized by most of his contemporaries: "He seemed to roam with no object in view, / Like One who had nothing on earth to do." But her message was that even the poorest and weakest souls could act to improve the world:

> Poor Johnny was bended well nigh double
> With years of toil, and care, and trouble;
> But his large old heart still felt the need
> Of doing for others some kindly deed.
>
> "But what can I do?" old Johnny said;
> "I who work so hard for daily bread?
> It takes heaps of money to do much good;
> I am far too poor to do as I would."

And having borne the taunts of those who had deemed her mad for her commitments and views, she imagined that Apple-Seed John experienced the same:

> In cities, some said the old man was crazy;
> While others said he was only lazy;
> But he took no notice of gibes and jeers,
> He knew he was working for future years.
> . . . .
> So he kept on traveling far and wide,
> Till his old limbs failed him, and he died.
> He said at last, " 'Tis a comfort to feel
> I've done good in the world, though not a great deal.[35]

The Johnny Appleseed myth that emerged in the years after his death became part of the national origin story. And that story was essentially a celebration of American empire—the transformation of a continent from savage to civilized. But mythmakers had to address the reality that their America was the result of the dispossession of the original inhabitants of the continent. For many nineteenth-century Americans, this reality did not trouble them in the least. They saw Native Americans as savages unworthy of compassion. The Indian-fighting exploits of Davy Crockett and Daniel Boone were retold without ambivalence or recoil. James Fenimore Cooper's bloody stories of American creation and dozens of dime store novel imitators sold briskly. Johnny Appleseed, too, became part of this celebration of the nation's origins, but his story was free from the taint of that violence. As such, he became, for sentimental reformers who were unsettled by the vulgarity and violence of the Crockett myths, a more appealing champion. Johnny Appleseed, after all, was a man of extraordinary gentleness and peace.

From the outset, then, Johnny Appleseed's relationship with Native Americans was a critical element of the story Northern reformers told. The Indians, one early fictional account insisted, "never molest[ed] him, but regarded him with a kind of superstitious veneration."[36] "By these wild and sanguinary savages he was regarded as a 'great medicine man,'" W. D. Haley wrote.[37] Another early chronicler asserted, "[N]o wonder the Indians liked him. They could read his character at a glance. All was revealed by his eye, clear as the sunlight of God."[38] The reality of John Chapman's relationship with Native Americans, of course, was more complex. No Native American account of John Chapman has survived. And for every general assertion by a white writer that Indians liked and trusted them, there is a more specific story that suggests this wasn't always the case. R. I. Curtis recalled John Chapman telling him stories of his "hair-breadth escapes" from Indians seeking to do him harm. W. Glines recounts a story of Indians stealing twenty ponies from Chapman. During the War of 1812, Chapman clearly sided with his white countrymen and on several occasions rushed to warn them of real or imagined Indian threats. And in later years he may have served as a bill collector for a white merchant hoping to recover debts from Ohio's last remaining Delaware.[39] Despite these facts, the vision of Johnny Appleseed as a unifying figure, beloved by both Indians and whites, continued to grow in its importance as Johnny Appleseed became a central part of the American origin myth. For white reformers, Johnny Appleseed, who befriended Indians and remade the continent with a hoe, not a rifle, was a more palatable hero.

But despite their humanitarian feeling for Native Americans, Northern reformers shared the almost universal assumption of white cultural superiority.

Not surprisingly, Johnny Appleseed emerged in the myth as a man who tried to lift Native Americans out of their savage condition. One early version of the myth that first appeared in the *German Reformed Messenger* entitled "Sammy Appleseed," told of the tree planter setting his seeds in Indian villages and seeking promises from the inhabitants to care for them but in most cases failing to secure those promises from the ignorant inhabitants. When, a few years later the Indians had abandoned these villages, the new white settlers found Johnny's trees at the village sites and transplanted them to their own nurseries.[40] (Interestingly, the attribution of wild apple trees to the efforts of Johnny Appleseed persists to this day. When I traveled around the Midwest to gather material for this book, I regularly encountered local historians who claimed Johnny Appleseed must have planted the wild apple trees the first settlers of their town found when they arrived. It did not occur to them to consider that native peoples could have planted them.)

Child, who throughout her life expressed a sincere compassion for the plight of Native Americans and championed their rights, also incorporated Appleseed's friendship with Native Americans into her poem. Yet even she could not escape the assumption of white cultural superiority, suggesting that Johnny's Indian friends were incapable of understanding the greater benefit of his apple tree planting activities:

> Sometimes an Indian of sturdy limb
> Came striding along and walked with him;
> And he who had food shared with the other,
> As if he had met a hungry brother.
>
> When the Indian saw how the bag was filled,
> And looked at the holes that the white man drilled,
> He thought to himself 'twas a silly plan
> To be planting seed for some future man.

In the years after the publication of Child's "Apple-Seed John," the narrative of John Chapman as a reformer spreading Northern, evangelical, middle-class values proved quite resilient, and the theme of his friendship with Indians only grew stronger. In the late nineteenth and early twentieth century, virtually every county in Ohio produced a published history, and Johnny Appleseed stories made it into many of them. The writers of county histories borrowed liberally from the Rice, Haley, and Child accounts, and from each other. They also added bits of stories unique to each location.

Even as the scattered fragments of John Chapman's life were being fashioned into a coherent myth, changes were afoot in America's apple orchards. For the first eighty years of the nineteenth century, America witnessed an explosion of diversity in apple varieties, as migration and markets colluded to produce new accidental varieties from the seedling orchards of frontier squatters and from the volunteer trees that had sprung up from seeds discarded in fields and hedgerows. Whenever a fruit of exceptional quality was discovered, it was propagated locally by grafting or rooting suckers. Some of these, such as Zebulon Gillette's Rome Beauty, found wider markets through commercial nurseries popping up in every state in the decades before the Civil War.[41] Attempts to identify and catalog every variety were futile. The first edition of Andrew Jackson Downing's *Fruits and Fruit Trees of America*, published in 1845, ran to 594 pages, but critics soon noted that while it contained most varieties known in the Hudson Valley and mid-Atlantic, it neglected many new fruits from the South and the West.[42] American orchards before 1880 were ubiquitous. The majority of farms contained an orchard of apple trees, except in places where climate made them unviable. These orchards were not the primary focus of a farmer's efforts; rather, they were one aspect of a farm that produced a variety of agricultural goods. These nineteenth-century orchards contained a radical diversity in fruit types, many of them peculiar to their locality or region.

But signs of trouble loomed. The proliferation of fruit orchards across the nation had created perfect environments for the propagation of pests that preyed upon them. After 1880 several factors altered the American orchard permanently. Among these was the invention of chemical treatments, the formation of the U.S. Department of Agriculture and national railroad system, and the invention of refrigerated box cars.

APPLES AND APPLESEED, 1880–1945

The first chemicals applied to apple orchards to fight pests arrived about 1880 from France, where horticulturalists had discovered that a highly toxic, bright-green powder made from ferrous sulfate and dubbed Paris Green was quite effective at controlling the destructive codling moth. A few years later, some American orchardists adopted another French innovation, the use of hydrated copper sulfate, called Bordeaux Mixture, to treat mold on grapes and scab on apples. By the end of the decade, the federal government, through the U.S. Department of Agriculture, was an enthusiastic champion of the new chemical treatments and began actively promoting their use to farmers. The

1887 Hatch Act created federally funded agricultural experiment stations in every state, and controlling pests and diseases that were damaging crops was at the top of the agenda for these new stations. The USDA's message to farmers was that they needed to devote more time and resources to the care of orchards in order to keep them productive. This was a sermon small farmers had been hearing from the improvement-minded "book farmers" for decades, but the costs of equipment, materials, and time to apply pesticides forced many farms involved in the production of a variety of agricultural goods to make a choice. The small orchard as one part of a diversified family farm enterprise began disappearing, but the change was not overnight. As late as 1910, 46.8 percent of American farms still grew apples, and about 20 percent grew other tree fruit.[43] Still, after 1880, some farmers began pursuing commercial orchard operation exclusively. Fruit culture became a distinct form of farming, and many of the new commercial orchardists called themselves "growers" instead of "farmers."[44]

At the same time, railroads and refrigeration opened opportunities for large-scale commercial fruit growing on the West Coast. Apple orchards became a booming business in the Pacific Northwest, while a new crop, the navel orange, emerged as an important crop in the warmer climes of California. In a strange coincidence of history, Eliza Lovell Tibbets, a California resident born and raised in the Cincinnati Swedenborgian community during the days when John Chapman was connected to it, was the earliest champion of the navel orange in the state.[45] West Coast orchards also changed the way Americans encountered fruit. In the East, orchardists rolled oak barrels out into the orchard and filled them with fruit. Apples arrived at market in the oak barrel, and fruit at the bottom of the barrel was often bruised, crushed, and less than desirable. The scarcity of hardwood in the West led to the invention of small, softwood shipping crates for packing apples. Soon crates of western apples, carried across the nation in refrigerated rail cars and decorated with colorful shipping labels, appeared in Eastern markets, and Western growers could boast that the fruit on the bottom of their modest-sized crates did not suffer the abuse of the apple barrel.[46] The emergence of a national commercial fruit industry had other effects as well. Growers began concentrating their efforts on just a few varieties of fruit, selected for their hardiness, productivity, keeping qualities, and appearance. Local varieties began to disappear, and by the early twentieth century, Americans north, south, east, and west ate Baldwins, Ben Davises, Jonathans, Northern Spys, and a few other varieties.[47]

The nationalization of the apple market went hand in hand with the nationalization of the Johnny Appleseed myth. In the first decades of the twentieth century, as a new generation of storytellers and poets wrote their own versions of the American story, Johnny Appleseed made his appearance in most of them. There was perhaps no more benign symbol to celebrate the process of American empire-building. Still, John Chapman's story had its greatest appeal to reform-minded writers from New England and the Midwest. Illinois-born poet Vachel Lindsay, who became one of Johnny Appleseed's great champions, called Appleseed "a New England kind of saint."[48] Yet Lindsay, who had family roots in both North and South, was determined to nationalize the Johnny Appleseed myth, and in his poems Johnny Appleseed took his place alongside George Washington, Andrew Jackson, and Abraham Lincoln in the pantheon of national heroes. Lindsay at times appeared to aspire to be a twentieth-century version of Johnny Appleseed. He set off on several "tramping expeditions" across the nation with little money in his pocket. His desire was to encounter ordinary Americans and swap poems for food and shelter. Lindsay combined a deep patriotism with a concern for the poor and dispossessed. Traveling across the nation he gave recitals of his poetry in a frantic, populist style he called "High Vaudeville."[49] He voted the socialist ticket, embraced pacifism, and had utopian dreams for his nation. Johnny Appleseed was a natural and favorite subject of his poetry, as he embodied the values Lindsay sought to promote in the simple act of planting apple seeds:

> An angel in each apple
> that touched the forest mold
> A ballot-box in each apple
> A state capital in each apple
> Great high schools, great colleges
> All America in each apple
> Each red, rich round and bouncing moon
> That touched the forest mold.[50]

For Lindsay, Appleseed was an American hero worthy of more renown than Henry David Thoreau, for he was not just a dreamer but a doer. Lindsay also advanced the idea of Chapman as a unifier of white and Indian peoples. "The Indians worship him," he wrote. "Stuck holy feathers in his hair. / Hailed him with austere delight. / The orchard god was their guest through the night."[51]

When Rosemary and Stephen Vincent Benét made Johnny Appleseed a subject of a poem in their children's book, *A Book of Americans,* the theme of Indian veneration appeared again.

The stalking Indian,
The beast in its lair
Did not hurt
While he was there.

For they could tell,
As wild things can,
That Jonathan Chapman
Was God's own man.[52]

John Chapman continued to be a symbol of American reform values throughout the 1920s and 1930s. Social critic Lewis Mumford suggested that the young men employed by Roosevelt's Civilian Conservation Corps were carrying on Chapman's work: "These youngsters will not merely reforest our barren slopes and fight insect pests; they will plant trees along bare roads, for shade and beauty, pushing the trail of Johnny Appleseed beyond the Alleghenies; they will keep up our otherwise too costly parkways and help extend them further; they will clear out the rural slums, trim up the rundown edges of our landscape, and bring music, art, and personal beauty into parts of the country that are now ugly, infamous, and unfit for human habitation."[53]

Johnny Appleseed seemed to have a special attraction to artists whose patriotism had a socialist bent. Howard Fast, who emerged in the post–World War II years as a popular and prolific writer of historical fiction, made Johnny Appleseed the hero of one of his earliest young adult novels, *The Tall Hunter*, first published in 1942. In the novel, an enraged frontiersman sets off on a quest to recover his wife, who had been kidnapped by Indians. Along the way, he seeks vengeance against any Indian he encounters, until he meets Johnny Appleseed, "that kind, gentle stranger who planted seeds in the wilderness so that men would find fruit to eat." Appleseed "tried also to sow in Richard's bitter heart the seeds of understanding and forgiveness." In the end, Appleseed not only heals the hunter's heart but returns his wife to him, leaving "them both with a legacy of peace."[54]

Once again, Johnny Appleseed offered an alternative version of the birth of American empire, one that rejected, rather than celebrated, the bloodiness of that birth. Howard Fast was working for the Office of War Information when he published *The Tall Hunter*, and one year later he joined the Communist Party USA. With the emergence of the Cold War, it quickly became impossible in the United States to profess love of country and simultaneously embrace socialist ideals. Fast was called before the House Un-American Activities Committee in 1950. When he refused to divulge the names of persons who had contributed money to an orphanage for children of American veterans who fought for the socialists in the Spanish Civil War, he was imprisoned for three months. For years he was blacklisted by publishers, but he continued to produce works of historical fiction that were highly patriotic in their message. Today, several of Fast's novels about the American Revolution are staples in American high schools.[55] But in the early years of the Cold War, any portrayals of Johnny Appleseed that hoped to get attention would have to tone down suggestions that the apple tree planter was a social radical.

APPLES AND APPLESEED IN THE LATE TWENTIETH CENTURY

After World War II, as Howard Fast discovered, the range of acceptable national myth narratives narrowed considerably. As the United States increasingly defined itself against Soviet communism, interpretations of Johnny Appleseed reflected this change. When Disney released an animated version of the Johnny Appleseed story in 1948, John's faith in God was front and center. The narrator states that three other great nation builders had their distinctive tools in their mission—Paul Bunyan had his hammer, John Henry his axe, and Davy Crockett his rifle—but Johnny Appleseed's tools were his bag of apple seeds and his Holy Bible. The cartoon opens with a young Johnny singing a Disney-created song that has come to be known as "The Johnny Appleseed Grace," and many believe it was actually written by Chapman.

> The Lord's been good to me
> And so I thank the Lord
> For giving me the things I need
> The sun and rain and the apple seed
> Yes He's been good to me.[56]

The Johnny Appleseed story told by Disney is a near perfect sermon on postwar American values. Faith in God and the ability of the individual to make a difference in history are the central themes. Johnny celebrates American freedom, singing, "Here I am 'neath the blue blue sky, doing as I please," thanking God for that freedom. Soon his attention is drawn to a long train of Conestoga wagons pushing west, each containing a pioneer family. The wagon train has its own song celebrating American individualism:

> Get on a wagon, rolling west
> Out to the great unknown

> Get on a wagon rolling west
> Where you'll be left alone.
>
> . . . .
>
> The rivers may be wide
> The mountains may be tall
> But nothing stops the pioneer
> we're trailblazers all.

While John longs to join them, he believes he cannot—that he is too weak and too small, and doesn't own the gear he needs. Johnny's "private guardian angel," sent down from heaven, convinces him that all he needs is his faith, his Bible, and his apple seeds. Johnny sets out through a rugged wilderness, "a little man all alone, without no knife, without no gun," but to avoid the impression that Johnny is a pacifist, Disney included a scene where he imagines he is shouldering a rifle like the ones he saw the men on the Conestoga wagons hold, and another where he picks up a stick from the woods, and pretends to aim and shoot with it. Notably, the Indian makes only a minor appearance in Disney's Johnny Appleseed. Instead, Johnny Appleseed works to win over the trust of the forests animals, convincing them, by his kindness, of the benign nature of his mission to transform the wilderness. The Disney story ends with an image of an aged Johnny Appleseed atop a ridge, his shadow stretching across a transformed landscape of fields and orchards.

> This little man,
> he throwed his shadow clear across the land,
> across a hundred thousand miles square
> and in that shadow everywhere
> you'll find he left his blessings three
> love and faith and the apple tree.

Despite the story's celebration of individualism, Disney's Johnny Appleseed stopped short of praising difference in favor of conformity. Johnny Appleseed was a generic Christian in the story, not an apostle of unconventional Swedenborgianism. Johnny Appleseed could be an eccentric in postwar America, but the boundaries of that difference were increasingly constrained in a culture that valued conformity even as it professed to celebrate the power of the free individual.

Social reformers continued to embrace the Johnny Appleseed story in the Cold War years, but liberals in this era were careful to position themselves in opposition to communism. Novelist, agricultural reformer, and social commentator Louis Bromfield had first written about Johnny Appleseed during the depression years, recounting stories his great aunt had told him as a child about John Chapman's alleged visits to her family's farm. For Bromfield, Appleseed had always represented agrarian values of benevolence and thrift, but in an op-ed piece he published in the *Cincinnati Enquirer* in 1955, Bromfield compared Appleseed to St. Francis of Assisi and confessed that on some evenings on his central Ohio farm, he walked to a high ridge to "think about Johnny Appleseed for a while instead of the war mongering columnists, the dictators, the Communists, the shady turncoat politicians, the greedy materialists. I can't help thinking how much the world needs people like Johnny Appleseed and Saint Francis and how, in the last analysis, both were the highest manifestations both of Christianity and of civilization itself."[57]

In the Cold War era, interest in Johnny Appleseed became increasingly the domain of children's educators and civic boosters hoping to capitalize on Johnny Appleseed tourism. There was one significant exception, the publication of *Johnny Appleseed: Man and Myth* by Robert Price, a professor of English at a small central Ohio college. With the support of longtime Johnny Appleseed enthusiast and philanthropist Carlos Burr Dawes, Price devoted seventeen years to researching Chapman and produced the only scholarly work on him to that time. There had been plenty of other attempts at biography, but the draw of myth had consistently turned those efforts away from the search for the truth and toward storytelling. Price managed to gather almost all of the extant sources on John Chapman and to produce an authoritative biography of the man and survey of the myth. Among the surprising things Price brought to light were county land records revealing that Chapman had purchased, sold, and defaulted on a significant amount of property during his adult years. That new information led to the creation of a new element of the Johnny Appleseed myth, Chapman as the successful businessman. This interpretation remains quite popular. A writer in *American Fruit Grower* magazine declared in 1981, at the dawn of the Reagan era, that "Johnny Appleseed was an entrepreneur—the kind of small businessman so much a part of the building of America—who conceived and executed a unique and daring enterprise of growing and selling apple tree seedlings... One of the key ingredients of business success is a sound understanding of the nature of the market served. Johnny Appleseed seems to have comprehended his market exactly." The writer disputed the idea that Chapman ever would have given his trees away, insisting that he "deliberately, and in a business-like way sold the seedlings to pioneer farmers."[58]

Such a characterization no doubt had appeal in an America where the businessman's reputation was more often than not ascendant. Of course, the measurement of success is a subjective one, and a close look at his overall record of land deals calls that into question. Furthermore, the very term "businessman" has become freighted with meanings that make the description of Chapman as a businessman problematic. Yet it remained a quite common metaphor for people to grasp onto once they realize that Chapman did not give all of his trees away for free and that he had acquired landholdings beyond what he could farm.

In the last half of the twentieth century, Johnny Appleseed remained a staple element of elementary school curriculums, especially in New England and the Midwest. New children's books appeared about him almost annually. And many communities in New England, Pennsylvania, Ohio, and Indiana that were connected to Chapman built monuments and established annual Johnny Appleseed festivals. By the 1960s, especially in the central Ohio communities where Chapman resided, Johnny Appleseed coffee shops, restaurants, shopping malls, and even one Johnny Appleseed Junior High sprung up.

While the Johnny Appleseed myth underwent a homogenization to make it safe for Cold War America, the American apple orchard received a postwar makeover as well. The critical innovations in the era were the emergence of high-density dwarfing stock orchards and technological advances that enhanced the appearance and keeping qualities of fruit. By the 1980s, most American apples grew on the cloned rootstocks of just one or two parent trees. These cloned dwarf rootstocks produced very small trees that provided an abundance of fruit in a tiny space. In these new orchards, tree density per acre increased five or six times, from roughly 40 or 50 trees per acre to a new standard of 240 cloned dwarf trees per acre.[59] The introduction of low-oxygen storage facilities meant that apples could be kept looking fresh for up to ten months after picking. And the development of new chemical growth regulators such as Alar allowed growers to pick uniformly dark red—but not quite ripe—apples from the tree for long-term storage. These new technologies raised the cost of production, forcing most growers to expand their operations to achieve greater economies of scale. Many unwilling or unable to expand their orchards shut down their operations or sold them to others. As commercial orchards grew in size, they did not generally grow in diversity, instead focusing production on a few reliable varieties.[60]

The postwar years witnessed the hegemony of the year-round Red Delicious apple in the grocery store. The Delicious was not a new apple. It was discovered on an Iowa farm back in 1879. Noted for its taste and its keeping qualities, the

In 1959, Shell Chemical used the image of Johnny Appleseed to champion the beneficial effects of their ammonia fertilizers on the American apple crop, asserting that they "pick[ed] up where Johnny Appleseed left off." The modern grocery store apple bears little resemblance to the apples that grew on Chapman's seedling trees. Courtesy of the Johnny Appleseed Society Museum, Urbana University, Urbana, Ohio.

Delicious became a favorite apple of growers. By the 1950s, nurseries interested in the physical appearance of apples developed several red sport versions of older varieties, and the Red Delicious proved to be the most successful of these. With its uniform deep red skin and its distinctive heart shape, the Red Delicious soon became the American apple of choice. Yet in the race to produce ever redder and redder specimens of perfectly shaped Red Delicious apples, growers abandoned the pursuit of flavor for appearance. By 1980 almost half of all apples sold in grocery stores were Red Delicious, with its cousin the Golden Delicious a close runner-up. And through low-oxygen storage methods, growers had conquered the seasons, making these visually perfect apple specimens available twelve months of the year. So long as consumers prized appearance and consistency above all other values, the Red Delicious would retain its dominant position.[61]

Commercial orchardists poured all their resources into producing perfect red apples. However, the Red Delicious and another popular red variety, the McIntosh, when grown on cloned dwarfing rootstock had a tendency to drop from the trees before they had achieved their full blush. To combat this, growers began applying a chemical with the commercial name of Alar to red apple varieties to slow their growth and prevent early drop. Tests on Alar had revealed that at very high doses it was a known carcinogen. The Environmental Protection Agency had not banned it, however, believing that it presented an extremely low risk to consumers. But Alar attracted the attention of environmental groups, most notably the National Resources Defense Council (NRDC), which lobbied for its ban, claiming the risk was far greater than the industry believed and that it was an absolutely unnecessary chemical. It was used for purely cosmetic purposes, not to protect the apples from a serious pest or disease.[62]

The story exploded when the investigative news show *60 Minutes* ran a story on Alar in February 1989, introduced with an illustration of a skull and crossbones superimposed on a Red Delicious apple. The apple, long a symbol of wholesomeness and health, was transformed overnight into a risk to families and children. Actress Meryl Streep, then a mother of a one-year-old, became the public face of an organization called Mothers and Others, which ran a campaign to press for Alar's ban. The market for apples temporarily crashed, and American growers took heavy losses on their crops.[63] Stemilt Growers, one of the largest in Washington State, responded with a "Responsible Choice" campaign later in the year and added the ladybug—a symbol of natural pest control—to their logo in an effort to reassure Americans that the apples they grew and marketed were safe.[64] Uniroyal Chemical, Alar's manufacturer, realizing it could not possibly find a future market for its product in the United States, withdrew it from the market before a legislative ban could be enacted.[65]

The Alar controversy proved to be a watershed moment in the story of the apple, and to this day, two competing narratives of its meaning vie for the public memory. For the fruit and agrochemical industries, the "Alar Scare" was a sinister manipulation of public fear by an environmental group that launched a savvy marketing campaign intended to destroy them, enlisting the liberal media and Hollywood star power to advance its own agenda. They point to subsequent studies that show the NRDC's initial assessment of risk to be exaggerated and claim that the result was perhaps hundreds of millions of dollars in lost revenue for growers and the collapse of some family farms. The story of anti-Alar hysteria in the wake of the *60 Minutes* piece produced its own urban legends. An unsourced story about a panicked mother who called the police to ask them to stop her child's school bus so she could remove the deadly apple from his lunchbox has been repeated ever since. Environmentalists respond by pointing out that foods with traces of potential carcinogens pose a greater risk to children than to adults and that the apple industry's lawsuits against *60 Minutes* and the NRDC resulted in victories for the defendants. Both sides have employed the story to advance their own cause. In 1996 environmentalists were able to press Congress into new regulations of agricultural chemicals in part as a result of the concerns about Alar, while the farm industry has been able to use the story to pass food defamation laws in many states—laws making it easier for agricultural producers to sue the media and celebrities for inciting scares about food safety that cost farmers money.[66]

But while the Alar controversy brought on a sudden, intense, and momentary crisis for American apple growers, a slower-developing but more significant threat had begun in the mid-1980s and reached its peak a decade later. Just as American consumers began to grow tired of the bland flavors and lack of variety of the Red and Golden Delicious monopoly, challengers from abroad entered the market with apples with fresh flavors and colors. Braeburns and Galas from New Zealand, Granny Smiths from Australia, and Fujis from Japan began appearing in American supermarkets. Apples from Chile were not far behind, and the result was that American consumers could now select apples in the spring that were fresh from the trees rather than from controlled atmosphere storage facilities.[67] Global competition proved to be a much tougher challenge to the American apple industry than concerns about chemicals. About 20 percent of American orchards went under in the 1980s and 1990s as a result.[68]

Small family orchards were the most vulnerable. Since 1945 there was a growing sense among some agricultural policy makers that the family farm was too labor-inefficient to compete in an increasingly globalizing food market. Earl Butz, secretary of agriculture under Nixon and Ford, warned American farmers to "get big or get out" and rewrote the rules of federal farm policy to encourage larger operations. Larger farms and orchards, of course, meant higher labor needs during harvest season, requiring the hiring of large numbers of seasonal migrant workers. The nineteenth-century family-farm model, which had long depended on the labor of children, perhaps a few hired hands, and the cooperation of neighbors during peak labor demand, was no longer viable. Following Butz's advice, by the 1980s and 1990s, most apple growers either got big or got out. Yet the idea of the family farm has run deep in the American psyche since Thomas Jefferson identified it as the locus of American virtue. In the 1980s, Willie Nelson and John Mellencamp, musicians with roots in rural America, began holding annual Farm Aid concerts to "save" the family farm and to raise awareness of the debt crisis family farmers faced.

But charity concerts were no match for the forces of global capitalism. Growing apples for the market became almost exclusively the domain of mega-growers. By 1997, just 1.5 percent of American farms still produced apples. Fruit growing was thoroughly specialized and dominated by big players. Large fruit farms, those with sales of over $1 million a year, represented just 3.8 percent of fruit-growing farms but produced almost 60 percent of the nation's fruit. The average large fruit farm in 1997 generated about $2.3 million in annual sales; for the rest, the average annual sales were about $63,000.[69] Among apple growers, Washington State emerged as the center of large-scale operations.

Some former family farms, like Stemilt in Wenatchee Washington, had begun to go down the "go big" road in the early 1960s, building their own packing houses and controlled atmosphere storage units, and buying up smaller struggling orchards. The American apple industry has also come to an accommodation with the free trade movement to a certain extent. Growers like Stemilt have taken the lessons of neoclassical economics to the next step and pursued a strategy of vertical integration—acquiring a stake in the apple import business and integrating southern hemisphere apples into their retail distribution networks from March to July, thereby sharing in the profits of imported fruit. Furthermore, American apple growers reaped some unexpected benefits from free trade agreements, as about one quarter of the American fresh apple crop is now exported to countries like Mexico and Columbia.[70]

Represented by lobbying organization U.S. Apple, America's mega-growers push for legislative action that is a mix of Smithian economics and statist protection. They support legislation that protects their access to flexible, low-wage seasonal labor during the picking and packing season and oppose health care and union rights legislation that threaten to increase their seasonal labor costs. They have sought to limit and control regulatory processes relating to the environment and consumer health and have opposed global warming legislation, which they believe threatens to increase their energy costs and make them less competitive in global markets—markets that increasingly shift production of all goods to the lowest-cost sites of production. At the same time, they have encouraged continued trade restrictions on fresh apple imports from their greatest potential threat—China—which currently produces more than half the world's apples and does so at costs the U.S. industry could not possibly match. Furthermore, U.S. Apple continues to lobby for direct and indirect subsidies by ensuring apples and apple products are included in government programs designed to promote children's nutrition and federal support for biotechnology and other food research that promises to increase yields and reduce costs.[71] In sum, the American apple industry has embraced what have become the standard rules of operation of multinational corporations in the twenty-first century—increasing inputs of technology and energy; reducing labor, regulatory, and other production costs; seeking vertical integration of production from tree to grocery store bin; and lobbying for favorable trade legislation and government subsidy where possible. Still, the romance of the family farm remains powerful in the United States, and a survey of the Web sites of major apple-growing corporations reveals that most of these giants emphasize their family farm roots.

American apple growers also have invested millions in marketing and advertising campaigns that seek to protect the apple's image as a symbol of health. In one of the most innovative efforts to promote the apple as a health food, Washington Apple growers formed a partnership with Gold's Gym to promote a diet book, *The 3-Apple-a-Day Plan: Your Foundation for Permanent Weight Loss*. Gold's Gym launched a twelve-week "Get-in-Shape Contest," in which contestants were encouraged to eat an apple before each meal. The company also declared the Washington Apple "the official diet pill of Gold's Gym."[72] Others have appealed to the American taste for the enhanced flavors of processed foods. Gräpple® now markets a Fuji apple that arrives in the grocery store in blister packaging. It has been saturated in a batch of flavorings so that it tastes like Concord Grapes.

Many of the new orchards emerging in response to global competition are grown on cloned stock so dwarfed that it cannot hold itself up and instead depend on trellises and stakes for support. These cloned dwarfs were planted with the expectation of ten-year lifespans, providing fruit in all but the first of the ten years. By the end of those ten years, the grower expects to tear them out and replace them with new varieties developed to excite and reawaken consumer demand.[73] Increasingly, these new varieties are privately owned by their developers, who license the rights to grow them but retain the authority to dictate to the grower how many trees he can grow, under what conditions he can grow them, and when he must tear them up. A revolution in the idea of what constitutes property in fruit trees no less significant than the one that saw Native American concepts fall to European ones is currently under way. Private varieties like Pink Lady™, Zestar!®, and Pinata™ are leading the way.[74]

A decade of turmoil and transition for the apple industry, the 1990s also proved to be one with a modest growth in interest in Johnny Appleseed. John Chapman found some important new champions in Ohio. Cincinnati resident William Ellery Jones became one of Chapman's most enthusiastic promoters and launched an ambitious plan to establish a Johnny Appleseed Heritage Center and Outdoor Drama near Mansfield. Jones toured the state telling Chapman's story and was successful in raising money for the Heritage Center and commissioning a play recounting Chapman's life. The drama opened in the summer of 2004 but closed after just two years of performances, not attracting large enough audiences to sustain it. Jones and the Heritage Center nonprofit are still working to bring it back.[75] Also in the late 1990s, faculty at Urbana University, the only existing undergraduate institution in the nation founded by the New Church, launched a Johnny Appleseed Society and established a museum on campus. That venture is still thriving, and the group has many active members. The society acquired the copyright to Robert Price's *Man and Myth* from Carlos Burr Dawes, the philanthropist who funded Price's efforts, and in 2000 they released a new edition, which quickly sold out. Jones and the Johnny Appleseed Society share a common vision of the uses of the Johnny Appleseed myth, seeing him as a hero and an excellent role model for children. They present a multifaceted, but generally saintlike, version of Chapman—successful businessman, environmentalist, philanthropist, and man of peace and great faith.[76]

APPLES AND APPLESEED IN THE TWENTY-FIRST CENTURY

With the dawn of the twenty-first century a handful of interpreters of the Johnny Appleseed legend have put forth some contrarian viewpoints. Michael

Pollan wrote extensively about John Chapman in his *Botany of Desire: A Plant's Eye View of the World*. Pollan toured many of the locations associated with Chapman with William Ellery Jones as his host but ultimately found Jones's vision of John Chapman to be too unbelievably saintly. Pollan criticized Jones for never mentioning that seedling apples were used on the frontier for making alcoholic cider and instead offered his own take on Chapman as "Dionysus's American Son." Chapman was for Pollan "a figure of the fluid margins, slipping back and forth between the realms of wildness and civilization, man and woman, man and god, beast and man." Pollan declared inaccurately that "just about the only reason to plant an orchard of the sort of seedling apples John Chapman had for sale would have been its intoxicating harvest of drink."[77] But he was right to point out that since alcoholic cider has disappeared from the American table, the apple had emerged in the twentieth century as an unambiguous symbol of health and wholesomeness and the loss of memory of the apple's diverse past had played a significant role in sanctifying and homogenizing the image of Johnny Appleseed for a modern American audience. For Pollan, Bill Jones's Johnny Appleseed is as bland as a Red Delicious, and he champions a more complicated rendering.

Pollan has not been alone in his efforts to upend the portrayal of John Chapman as a frontier saint. A few other voices have joined the chorus. Agricultural historian David Diamond has taken umbrage at the exaggerated credit Chapman receives for populating the nation with apple trees. "Appleseed fails in many ways to personify the systematic industry of late colonial, revolutionary, or antebellum farmers and orchardists—defined by stable family units, agricultural diligence, craft self-reliance, and civic engagement," Diamond insists. "Indeed, it is remarkable that the unwed, itinerant Chapman, who avoided farming and was perpetually dependent on his married sister would be considered to be a planter of crops at all."[78] Diamond is correct in his assertion that the myth greatly exaggerates the reach of Chapman's nurseries but invents his own myth with the allegation that Chapman was perpetually dependent on a married sister for support. Diamond's sympathies lie with the improving horticulturists who spread grafted fruit to the West, and he appears to have adopted the prejudices of some of that class in his disdain for Chapman's lifestyle and his seedling orchards. Finally, in 2011 two new accounts of Chapman's life were published. A new biography by journalist Howard Means concluded that "by our modern definitions, John Chapman almost certainly was insane."[79] In contrast, Robert Morgan included John Chapman as one of ten profiles in *Lions of the West: Heroes and Villains of the Westward Expansion*. Morgan declares Johnny

Appleseed to be the "the saint of western frontier folklore [and] Crockett . . . its martyr." For Morgan, Johnny Appleseed embodies American freedom but a freedom quite different from that put forward in the early Cold War era, "an ultimate freedom to just be yourself, unconcerned with status, fashion, excess possessions or comforts." An early reviewer of Morgan's work, however, protested John Chapman's inclusion among the "Lions of the West," asserting that Chapman was not representative of the expansionist spirit and rejecting the idea that he played a significant role in western expansion.[80]

Whether Pollan, Diamond, Means, or Morgan's critical reviewer make any headway in upending the myth of Johnny Appleseed is unlikely; Robert Price's methodical effort to sort "man" from "myth" has done little to draw clear lines in the public mind since its publication more than a half century ago. A compelling tall tale will always have more sticking power than a careful rendering of facts. But myths have their uses and their value. They teach us about ourselves, or at least the values we aspire to, and the story of the mythical Johnny Appleseed will outlive us all.

Nonetheless, assessing John Chapman the man, enrobed as he is in thick layers of tall tales and myths, is a challenging but worthwhile task. Were he easy to categorize and explain, he would not be remembered today. He survived in family stories precisely because he was peculiar enough to tell stories about. That does not mean he was not a product of his times. John Chapman was descended from English Puritans who came to America to live out their faiths, establish freehold farms, and pass a competency along to their heirs. By the time his father Nathaniel was born, this aspiration was just as strong, but harder to achieve. John grew up in the household of a poor New England carpenter-farmer who took up arms "in the pursuit of happiness," a happiness, when defined by the values of landholding independence, that continued to slip out of his grasp. John Chapman inherited no property from his father but instead inherited a revolution—a new world born of his father's sacrifices—and he set out as a young man to find his own place in this new world.[81] For all of his peculiarities, he had much in common with other Americans of his generation. He was at different times in his life a white man eking out an existence in the middle ground between Native American and European societies; a dirt poor squatter among squatters and land barons; and later a petty speculator among great and small speculators. He was a meek and somewhat effeminate man on a frontier first characterized by hypermasculinity and later a bachelor in a domesticated world reshaped by family cabins and farms. He was a Yankee peddler of sorts, a devotee of obscure religious doctrines at a time when many Americans embraced expanded religious freedoms and sought their own alternative versions of faith. He was an evangelist in a time when evangelism reached unprecedented levels of intensity and organization; he was a believer in a world of believers, anythingarians, and nothingarians. He was a monomaniacal planter of apple seedlings in the early years of American horticulture. He was a perfectionist and an idealist in an age filled with perfectionists and idealists.

Assessing the meaning of the apple, a fruit also enshrined as a cultural symbol, in American history is also no simple task. Its significance has also evolved with the nation. From its arrival with the first Europeans, the apple represented on this continent an alien conception of property. The ownership of a tree and the fruit that grew from its branches did not exist as an idea in North America before Europeans brought it here on the first ships. As Englishmen planted colonies along the East Coast of North America, they laid out orchards as both a symbol of their ownership and a sign of the "fruitfulness" of the land. From the outset, orchards delineated the line between rich and poor, as the former invested substantial resources in tidy, well-tended orchards and the latter appreciated God's gifts ripening each year in a neglected orchard that did not demand much of their scarce labor to yield rewards. The self-provisioning farmer's orchard that provided an abundance of household uses from cider and vinegar to dried apples and hog feed eventually yielded to the market-centered farmer's orchard of fruit destined for the cider mill, the brandy distillery, or the big city market as fresh fruit. The temperance movement's campaign against cider apples accelerated the shift to an age when Americans no longer drank their apples but ate them fresh instead. As the memory of hard cider faded, the apple emerged as a symbol of wholesomeness and good health, a reputation it retains, if in somewhat tarnished form, in an age concerned with pesticides. But the rise of the Red Delicious saw the apple's reputation gradually shift from the perfect fruit to the perfectly ordinary one, as consumer desires changed. The steady introduction of new patented, highly marketed varieties, like Pink Lady™ and Jazz™, reveal a new market savvy among America's apple growers, as well as radically new ideas about what it means to "own" fruit.

But the apple and Johnny Appleseed are once again emerging as powerful symbols pushing back against the transformations global industrial capitalism continues to bring. The increasing distance of farms and orchards from urban and suburban America, the disappearance of the labor-inefficient but idolized family farm, the relentless outward sprawl of cities into once agrarian hinterlands, and the dominance of a global agro-industry in the American kitchen face increasing resistance. In October 2000, when news spread around Leominster,

Massachusetts, the birthplace of John Chapman, that the town's last orchard was for sale and the land might be developed into a new housing subdivision, community members quickly organized to save it and played on the town's connection with Johnny Appleseed in their media campaign, suggesting that perhaps the young John Chapman had once wandered among apple trees on the very land under threat. Their campaign was a success, the housing development was defeated, and the land today remains an orchard, cooperatively managed by community volunteers under an organization called Friends of Sholan Farms.[82]

The fight to save the Sholan Farms, and its conversion into a community-run orchard, is just one manifestation of a "civic agriculture" movement that is as ambivalent about unrestrained capitalism today as many of John Chapman's neighbors were in his own time. Advocates of civic or sustainable agriculture call for a return to a farming that is local and fosters direct connections between producers and consumers. They reject the "get big or get out" mindset of global industrial capitalism, believing it is destructive of communities and personal health.

A 2004 documentary, *Broken Limbs* told the story of the Washington apple industry's crisis and the near extinction of small growers from the state. The film described not just the larger movement toward consolidation and corporatization of the state's apple orchards but celebrated a smaller counter-movement as well, the emergence of the sustainable farming movement among small apple growers. Dubbed the "New American Farmers" by sustainable agriculture champion Dr. John Ikerd, these small farmers survive in the shadows of the corporate giants by embracing a different set of priorities. They are, according to Ikerd, "[f]armers for whom money mattered but not above all else. What mattered most was holding on to a lifestyle they believe in. Because it was good for their families. Good for their communities. Good for their world." *Broken Limbs* showcased the stories of small producers like Grant Gibbs, who has succeeded with a strategy quite different from his corporate orchard neighbors: reduce costs, think small, focus on production of high-quality fruit sold directly to co-ops and health-oriented outlets, and set your own price, don't let the market set it for you. In the film Gibbs, described by the narrator as "a man on the fringes of society," declares, "I live a very modest lifestyle. I live within my means," and he boasts of his ingenuity for simplicity. His orchard consists of a range of heirloom varieties, mostly from the eighteenth and nineteenth century, which ripen at different times. As a result, he needs to own just one bin to pick them all.[83] Gibbs's genius at finding the alternative path to success contains echoes of John Chapman's ingenious frugality.

The apple has been embraced as a symbol of a better alternative world by others as well. Since the 1980s, a small but growing crowd of people has been recovering and replanting antique or heirloom varieties, old regional varieties popular in the days of John Chapman but discarded and almost lost when the industry went national in the late nineteenth century. Creighton Lee Calhoun published a catalog of disappearing *Old Southern Apples* in 1995.[84] The book sold out and commanded a very high price on the used book market before he released a revised and updated edition in 2011. Applesource an Indiana-based company founded in 1983 has found great success shipping sampler boxes of fresh-picked antique varieties directly to consumers each fall. Despite the fact that a box of twelve apples costs about $36 with expedited shipping, the company has not lacked for customers.[85] Interest in organically grown apples, sparked first by the Alar scare, continues to grow. While pesticide use and residue on commercial apples is more tightly regulated today than it was in 1989, in 2011 the apple earned the top spot on the Environmental Working Group's "Dirty Dozen"—twelve fruits and vegetables consumers should avoid or embrace organic, claiming a grocery store apple contained more pesticide residue today than any other nonorganic fruit or vegetable.[86]

Urban and peri-urban areas like Leominster have been the focus of the civic agriculture movement.[87] In big cities across America, volunteers are coming together to make fruit freely available to city dwellers. A Los Angeles–based organization, Fallen Fruit, publishes maps of fruit trees for different neighborhoods across the city.[88] An old city law protects the right of any city resident to pluck fruit from the branches of fruit trees that overhang sidewalks and public roadways, and Fallen Fruit encourages people to glean what they need but share with others. The Portland Fruit Tree project cares for city orchards and also enlists volunteers to help property owners harvest surplus fruit. Volunteers are permitted to take what they need, but much of the fruit is sent to food pantries and distributed to the poor.[89] Seattle has also launched a City Fruit program.[90] Yet the movement is growing beyond areas with large numbers of professionals willing to pay premium prices for quality fresh produce. Detroit, Michigan, one of the biggest losers in the march of global capitalism, now sees civic agriculture as an integral part of its efforts to turn itself around. Gardens and apple orchards are springing up in blocks only recently filled with abandoned houses. Urban apple trees are emerging as a powerful symbol of opposition to global capitalism and its impact on people and communities.

Perhaps no modern reformer represents modern apple idealism better than Lisa Gross who formed the Boston Tree Party in 2011 with the goal of planting a "decentralized urban apple orchard" to provide annually "thousands of free apples for all." Taking as its motto *frux civilas* (civic fruit), the enthusiastic volunteers of the Boston Tree Party began planting pairs of heirloom apple trees across Boston neighborhoods in the spring of 2011. The seal of the organization promises their mission will create interdependence, abundance, health, and cross-pollination. Apples, insists Gross, represent health and well-being; they are a "democratic fruit," and the heirloom varieties planted by the Boston Tree Party are an effort to recover a "loss of biodiversity, history, culture, and flavor." When Lisa Gross speaks, her belief in the power of apples to improve the world is evangelical: "Imagine our cities filled with fruit trees . . . planted in civic spaces, at schools and hospitals, parks and businesses, houses of worship and more. Imagine these communities coming together to care for these trees, to harvest and share their fruit. Imagine these trees as tools of environmental restoration, helping to restore the health of our soil, to improve air quality and to absorb rainwater runoff. Imagine these trees as community focal points, opportunities for participation, learning, and connection. This is the vision of the Boston Tree Party."[91]

It might just describe the vision of one of her Yankee forebears as well.

Notes

INTRODUCTION

1. The Cold War television version of Daniel Boone was not a Disney creation but instead developed by Twentieth Century Fox, which modeled him closely on Disney's Davy Crockett and employed the same actor, Fess Parker, in the title role.

2. For a discussion of the differences between biography and microhistory, see Jill Lepore, "Historians Who Love Too Much: Reflections on Microhistory and Biography," *Journal of American History* 88, no. 1 (June 2001): 129–44.

3. Barrie E. Juniper and David J. Mabberley, *The Story of the Apple* (Portland, Ore.: Timber Press, 2006), 17, 34–45; Frank Browning, *Apples: The Story of the Fruit of Temptation* (New York: North Point, 1998), 45–46.

4. Henry David Thoreau, *Wild Fruits: Thoreau's Rediscovered Last Manuscript*, ed. Bradley Dean (New York: W. W. Norton, 2000), 79.

CHAPTER 1: SEEDS

1. Jacob Chapman, *Edward Chapman of Ipswich, Mass., 1642–1678, and His Descendants* (Concord, N.H., 1893), 1–3; Virginia DeJohn Anderson, *New England's Generation: The Great Migration and the Formation of Society and Culture in the Seventeenth Century* (Cambridge: Cambridge University Press, 1993), 18–26.

2. Daniel Vickers, *Farmers and Fishermen: Two Centuries of Work in Essex County, 1630–1850* (Chapel Hill: University of North Carolina Press, 1994), 1.

3. Brian Donahue, *The Great Meadow: Farmers and the Land in Colonial Concord* (New Haven, Conn.: Yale University, 2004), 40; William Cronon, *Changes in the Land: Indians, Colonists and the Ecology of New England* (New York: Hill & Wang, 1983), 37–38.

4. Joseph B. Felt, *History of Ipswich, Essex, and Hamilton* (Cambridge, Mass., 1834); Florence E. Wheeler, "John Chapman's Line of Descent from Edward Chapman of Ipswich," *Ohio Archeological and Historical Quarterly* 48 (1939): 28–33, and "John Chapman Genealogy," Special Collections, Leominster Public Library, Leominster, Mass.

5. Kenneth Lockridge, *A New England Town: The First Hundred Years* (New York: W. W. Norton, 1985), described the New England town as a utopian communalist venture. Subsequent scholars have challenged this interpretation, including John Frederick Martin, *Profits in the Wilderness: Entrepreneurship and the Founding of New England Towns in the Seventeenth Century*

(Chapel Hill: University of North Carolina Press, 1991), who argues that towns were a capitalist, profit-minded venture from the outset. David Hall, *A Reforming People: Puritanism and the Transformation of Public Life in New England* (New York: Alfred A. Knopf, 2011), makes a convincing case that the Puritan founders were authentic reformers, out to create a better version of the England they left behind.

6. U. P. Hedrick, *A History of Horticulture in America to 1860* (New York: Oxford University Press, 1950), 30.

7. Donahue, *Great Meadow*, 165–66; Sarah F. McMahon, "A Comfortable Subsistence: The Changing Composition of Diet in Rural New England, 1620–1840," *William and Mary Quarterly*, 3rd ser., 42, no. 1 (Jan. 1985): 59. Donahue has found that the shift from beer to apple cider occurred in Concord almost from the outset of the town's establishment. In a study of widow's portions in Middlesex County across the seventeenth and early eighteenth century, Sarah McMahon found evidence that across the county, the shift to hard cider came in the eighteenth century. Both scholars agree that cider was ascendant in the eighteenth century, for reasons discussed later in this chapter.

8. Donahue, 165–66; Noel Kingsbury, *Hybrid: The History and Science of Plant Breeding* (Chicago: University of Chicago Press, 2009), 399–400.

9. Felt, *History of Ipswich*, 10–18.

10. Ibid., 19–21; Chapman, *Edward Chapman*, 2, 5.

11. Felt, *History of Ipswich*, 10–18; Chapman, *Edward Chapman*, 1–3, 5–7.

12. Hall, *Reforming People*, 149–52.

13. Chapman, *Edward Chapman*, 3, 5–6.

14. Ibid., 3.

15. Ibid., 113.

16. Ibid., 114.

17. Indians by no means disappeared in eastern Massachusetts after King Philip's War, but white officials used the war as justification to take control of almost all remaining Indian lands. Local historians in subsequent years wrote Indians out of the history of eastern Massachusetts after 1676. See Jean M. O'Brien, *Firsting and Lasting: Writing Indians out of Existence in New England* (Minneapolis: University of Minnesota Press, 2010), 34.

18. "Vital Records of Tewksbury, Massachusetts, to the End of 1849," http://ma-vital records.org/MA/Middlesex/Tewksbury/; Wheeler, "John Chapman's Line."

19. Vickers, *Farmers and Fishermen*, 256.

20. Wheeler, "John Chapman's Line."

21. Quoted in Rufus Phineas Stebbins, *A Centennial Discourse Delivered to the First Congregational Church and Society in Leominster, September 24, 1843* (Boston, 1843), 24.

22. Donahue, *Great Meadow*, 162–65; Hedrick, *History of Horticulture*, 32–33, John Henris, "Apples Abound: Farmers, Orchards, and the Cultural Landscapes of Agrarian Reform, 1820–1860" (Ph.D. diss., University of Akron, May 2009), 130.

23. Henry David Thoreau, *Wild Fruits* (New York: W. W. Norton, 2000), 92.

24. Robert Price, *Johnny Appleseed: Man and Myth* (Gloucester, Mass.: P. Smith, 1967), 12.

25. David Wilder, *History of Leominster* (Fitchburg, Mass., 1853), 42; John Shy, *A People Numerous and Armed* (Ann Arbor: University of Michigan Press, 1990), 37–38; Robert Gross, *Minutemen and their World* (New York: Hill & Wang, 1976), 70–71; James Kirby Martin and Mark

Edward Lender, *A Respectable Army: The Military Origins of the Republic* (Arlington Heights, Ill.: Harlan Davidson, 1982), 17–18.

26. Wilder, *History of Leominster*, 42; Wheeler, "John Chapman's Line," 29–30; Leominster First Congregational Church Records, Leominster Historical Society, Leominster, Mass., 32, 59.

27. Shy, *People Numerous*, 171–73.

28. Price, *Johnny Appleseed*, 13–14.

29. Henry S. Nourse, *The Military Annals of Lancaster, Massachusetts* (Lancaster, 1889), 112–13; Martin and Lender, *Respectable Army*, 53–55.

30. Price, *Johnny Appleseed*, 15.

31. William Kerrigan, "Apples on the Border: Orchards and the Conquest for the Great Lakes," *Michigan Historical Review* 34, no. 1 (2008): 25, 38–41.

32. D. S. Whittlesey, "The Springfield Armory" (Thesis, University of Chicago, 1920), 72–91.

33. *Journals of the Continental Congress* 11:472 and 12:670–72, 793. Much of this material was unearthed by George B. Huff and is included in a paper he wrote, "Let's Put the Record Straight," a copy of which resides in the Warren County, Penn., Historical Society Archives. Huff's interest was in clearing the reputation of Nathaniel Chapman, one of his ancestors, from Robert Price's suggestion that he was dishonorably discharged.

34. Wheeler, "John Chapman's Line," 30.

35. Richard Salter Storrs, "Genealogical Appendix," in *Proceedings at the Centennial Celebration of the incorporation of the Town of Longmeadow* (Longmeadow, Mass., 1884), 178–79; *Old Long Meddowe on the Hill* (Longmeadow, Mass.: Longmeadow Historical Society, 1982), 16–17.

36. Storrs, "Genealogical Appendix," 51.

37. Ibid.

38. "A Town List of the Quantity of Land with the Buildings Thereon and the Number of Inhabitants in the Town of Longmeadow ... Agreeable to an Act of the General Court, July 2, 1784," typescript, Longmeadow, Genealogy Room, Storrs Library, Longmeadow, Mass., n.p.

39. Wheeler, "John Chapman's Line."

40. Storrs, "Genealogical Appendix," 178–79; *Old Long Meddowe*, 16–17.

41. There are several excellent studies of politics and society in early Massachusetts towns. See, for example, Lockridge, *New England Town*, and Patricia Tracy, *Jonathan Edwards, Pastor: Religion and Society in Eighteenth-Century Northampton* (New York: Hill & Wang, 1980).

42. Linda M. Rodger and Mary S. Rogeness, eds., *Reflections of Longmeadow* (West Kennebunk, Maine: Longmeadow Historical Society, 1983), 1–2, 8–12.

43. For the full story, see John Demos, *The Unredeemed Captive: A Family Story from Early America* (New York: Alfred A. Knopf, 1994).

44. Storrs, "Genealogical Appendix," 228–29; Demos, *Unredeemed Captive*, 193.

45. *Old Long Meddowe*, 5–8; Demos, *Unredeemed Captive*, 194.

46. Rodger and Rogeness, *Reflections*, 12–14.

47. Ibid., 16.

48. Christopher Clark, *Roots of Rural Capitalism, Western Massachusetts, 1780–1860* (Ithaca, N.Y.: Cornell University Press, 1990), 44–45.

49. Leonard Richards, *Shays's Rebellion: The American Revolution's Final Battle* (Philadelphia: University of Pennsylvania Press, 2002), 81–83; Clark, *Roots*, 46–48.

50. Richards, *Shays's Rebellion*, 81–83.

51. "Town List," n.p.

52. This was certainly one of the two farms Nathaniel would later tell an Ohio neighbor he lost during the war years. Nathaniel and Lucy and their growing brood remained on the rented farm until 1800, when they picked up and moved west in the hopes of sunnier prospects, leaving a scarce trace of their twenty year presence in Longmeadow. Huff, "Let's Put the Record Straight," iv.

53. Quoted in Howard Zinn, *A People's History of the United States* (New York: Harper Perennial, 2010), 92.

54. Vickers, *Farmers and Fisherman*, 221–22. U.S. Federal Consensus, 1790 and 1800, Hampshire County, Longmeadow township.

55. For one example of the nature of bonded apprenticeship during these years, see Asa Sheldon, *Life of Asa G. Sheldon, Wilmington Farmer, in Two Arrangements* (Woburn, Mass., 1862), 23–38.

CHAPTER 2: BECOMING JOHNNY APPLESEED

1. Judge Lansing Wetmore's account, first delivered in a series of lectures on local history in 1853, is reprinted in J. S. Schenk, *History of Warren County* (Syracuse, N.Y., 1887) 153–54.

2. Wetmore dated the crossing as November 1797 (ibid.), but other sources place Chapman in the region fifteen months earlier: see Fort Franklin stores ledger, Sept. 3, 1796, Crawford County Historical Society, Meadville, Pa., 35.

3. Schenk, *History of Warren County*, 154.

4. Fort Franklin ledger, Sept. 3, 1796, 35.

5. For a full account of the contested history of the Wyoming Valley, see Paul B. Moyer, *Wild Yankees: The Struggle for Independence along Pennsylvania's Revolutionary Frontier* (Ithaca, N.Y.: Cornell University Press, 2007).

6. Ibid., 102–5.

7. Henry B. Plumb, *History of Hanover Township and the Wyoming Valley* (Wilkes-Barre, Pa., 1885), 237.

8. Ibid.

9. Paul A. W. Wallace, *Indian Paths of Pennsylvania* (Harrisburg: Pennsylvania Historical and Museum Commission, 2005), 46–48, 66–69.

10. Ibid., 3–4.

11. See account entries for John Chapman in Fort Franklin ledgers, 35, 46.

12. Wallace, *Indian Paths*, 8.

13. Alfred Huidekoper, "Incidents in the Early History of Crawford County, Pennsylvania," in *Memoirs of the Historical Society of Pennsylvania*, ed. Henry C. Baird (Philadelphia, 1850) 4:93–94.

14. Alexander McDowell, "Deposition," in *Report of the Case of the Commonwealth vs. Tench Coxe, Esq.* (Philadelphia, 1803), 43–45.

15. "Autobiography of Cornelius Van Horn" (1837), 12–13, unpublished oral history, Crawford County Historical Society, Meadville, Pa.

16. Robert Price, *Johnny Appleseed: Man and Myth* (Gloucester, Mass.: P. Smith, 1967), 21, 23; George Burges, *A Journal of a Surveying Trip into Western Pennsylvania . . . in the Year 1795* (Mt. Pleasant, Mich.: John Cumming, 1965), 34–35.

17. William H. Egle, *An Illustrated History of the Commonwealth of Pennsylvania* (Harrisburg, 1876), 1134.

18. U.S. Federal Census, 1800, 1810, Warren County, Pa.

19. Anthony F. C. Wallace, *The Death and Rebirth of the Seneca* (New York: Vintage, 1969), 187–88; Thomas Abler, *Cornplanter: Chief Warrior of the Allegany Senecas* (Syracuse, N.Y.: Syracuse University Press, 2007), 135–37.

20. Glenn F. Williams, *Year of the Hangman: George Washington's Campaign against the Iroquois* (Yardley, Pa.: Westholme, 2007), 256.

21. See Abler, *Cornplanter*, 58–84, for a summary of Cornplanter's diplomacy after the Revolution.

22. Wallace, *Death and Rebirth*, 187–88; Abler, *Cornplanter*, 135–37.

23. David Swatzler, *A Friend among the Senecas: The Quaker Mission to Cornplanter's People* (Mechanicsburg, Pa.: Stackpole, 2000), 474–8, Abler, *Cornplanter*, 138.

24. Abler, *Cornplanter*, 136, Wallace, *Death and Rebirth*, 192.

25. Wallace, *Death and Rebirth*, 184. All of Warren County counted only 233 white residents in 1800; U.S. Federal Census, 1800, Warren County.

26. Swatzler, *Friend among the Senecas*, 149–52. Dean R. Snow, *The Iroquois* (Malden, Mass.: Blackwell, 1996), 108.

27. Swatzler, *Friend among the Senecas*, 151.

28. R. I. Curtis, "John Chapman, alias 'Johnny Appleseed,'" *Ohio Pomological Society Transactions* (Columbus, Ohio, 1859), 68–69.

29. See Richard White's *The Middle Ground: Indians, Empires, and Republics in the Great Lakes Region, 1650–1815* (New York: Cambridge University Press, 1991), xi–xv.

30. John Daniels ledger, Warren County Historical Society, Warren, Pa.

31. David Thomas, *Travels through the Western Country in 1816* (Auburn, N.Y., 1819), 43–44.

32. Curtis, "John Chapman," 68–69.

33. On the ecological consequences of deforestation, see William Cronon, *Changes in the Land: Indians, Colonists, and the Ecology of New England* (New York: Hill & Wang, 1983), 122–26.

34. Daniels ledger, 39; Curtis, "John Chapman," 68–69.

35. Swatzler, *Friend among the Seneca*, 27–54; Matthew Dennis, *Seneca Possessed* (Philadelphia: University of Pennsylvania Press, 2010), 119–20, 151–52.

36. J. Hector St. John de Crèvecoeur, *Letters from an American Farmer and Sketches of Eighteenth-Century America*, ed. Albert E. Stone, (New York: Penguin, 1986), 220.

37. Fort Franklin ledger, Aug. 6, 1797, 46.

38. Daniels ledger, 39.

39. Charles E. Williams, "What Was the 'Broken Straw' of Pennsylvania's Brokenstraw Creek?—An Ethnobotanical Inquiry," presented at the Society of Ethnobiology, Thirty-Fourth Annual Meeting, Columbus, Ohio, May 5, 2011.

40. Paul D. Evans, *The Holland Land Company* (1924; reprint, Fairfield, N.J.: Augustus M. Kelley, 1979), 107–10.

41. *Report of the Case*, 4–6.

42. Norman B. Wilkinson, *Land Policy and Speculation in Pennsylvania* (New York: Arno, 1979), 129–31; Terry Bouton, *Taming Democracy: The People, the Founders, and the Troubled Ending of the American Revolution* (New York: Oxford University Press), 222–23.

43. Robert D. Arbuckle, *Pennsylvania Speculator and Patriot* (University Park: Pennsylvania State University Press, 1975), 76.

44. Ibid., 28, 76; "Gen'l William Irvine to Pres. Dickinson—Donations Lands, 1785," in *Pennsylvania Archives*, series 1, 11:512–20.

45. Two provision accounts books for the Holland Land Company kept by Alexander McDowell are in the collections of the Crawford County Historical Society, Meadville, Pa. They testify to the enormous sums the company expended in an effort to entice and aid settlers. See also the appendix to *Report of the Case*, 56–57.

46. The full text of the land law can be found in *Report of the Case*, 38–43.

47. Ibid., 56–58.

48. Nicholas B. Wainwright, *The Irvine Story* (Philadelphia: Historical Society of Pennsylvania, 1964), 2–11.

49. Ibid., 12–14.

50. Callender to William Irvine, May 17, 1797, Newbold-Irvine Papers, Historical Society of Pennsylvania, Philadelphia.

51. William to Callender Irvine, May 26, June 12, June 25, June 27, 1797; Callender to William Irvine, May 17, June 13, 1797, Newbold-Irvine Papers.

52. Wainwright, *Irvine Story*, 17.

53. Huidekoper, "Incidents," 106.

54. Matt Young to Callender Irvine, June 12, 1798, and May 27, 1799; Callender Irvine to Matt Young, n.d., Newbold-Irvine Papers.

55. Matt Young to Callender Irvine, May 27, 1799. Newbold-Irvine Papers.

56. Unknown to Callender Irvine, Dec. 7, 1799. Newbold-Irvine Papers.

57. Callender Irvine to William Irvine, Presque Isle, June 27, 1804; Robert Andrews to Callender Irvine, July 13, 1804, Newbold-Irvine Papers.

58. J. H. Newton, *History of Venango County* (Columbus, Ohio: J. A. Caldwell, 1879), 595.

59. Ibid., 526.

60. Most of the Holland Land Company records for northwestern Pennsylvania were carted back to the Netherlands in a disordered trunk when the company finally pulled up stakes. The best study of the Holland Land Company that makes some use of those records is Evans, *Holland Land Company*.

61. The store ledger at Franklin contains entries for John Chapman in September and November 1796 and February 1797; the Daniels ledger for the store at Brokenstraw testifies to his presence in the region in March and May 1797. In July, August, September, and October 1797 he was once again listed in the Franklin ledger. The next entry for John Chapman was at the Brokenstraw store in May 1798.

62. McDowell, provision accounts book, 599.

63. E. L. Babbitt, *Allegheny Pilot* (Freeport, Pa., 1855), 6–7.

64. Daniels ledger, 121; Fort Franklin ledger, 35, 40, 46.

65. A Pittsburgh oral tradition tells of John's arrival with his younger brother in 1792, but no corroborating evidence survives of John or Nathaniel's presence in Pittsburgh, and the two boys would have been just seventeen and eleven in that year.

66. Nathaniel's appearance in the Brokenstraw ledger was limited to the summer of 1798, and no additional documentary traces of his presence in the region have been uncovered.

His adventures on the Allegheny plateau may have been limited to a solitary summer, and his older brother may have led him back to Longmeadow in the fall. Nathaniel and most of the rest of John's family migrated to the Marietta, Ohio, region a few years later.

67. "Autobiography of Cornelius Van Horn," 17.

68. George Powers store ledger, Venango County Historical Society Special Collections, Franklin, Pa.

69. Burges, *Journal of a Surveying Trip*, 10.

70. Irvine to Dickinson, Aug. 17, 1785, in *Pennsylvania Archives*, Series 1, 11:515.

71. U.S. Federal Census, 1800, Allegheny, Franklin, and Irwin Townships, Venango County, Pa.

72. Fort Franklin ledger, 46.

73. Curtis, "John Chapman," 68–69. Curtis makes a brief reference to the ice floe story. A fuller account comes from another childhood memoir of Chapman, by W. M. Glines, who encountered him in Marietta, Ohio; W. M. Glines, *Johnny Appleseed by One Who Knew Him* (Columbus, Ohio: F. J. Heer, 1922).

74. Curtis, "John Chapman," 69.

75. Chapman's bills of exchange are transcribed in "Estate Papers of John Chapman," Ohio Historical Society, Columbus. See also J. A. Caldwell, *History of Belmont and Jefferson Counties* (Wheeling, W. Va., 1880), 191–92.

76. Babbitt, *Allegheny Pilot*, 6–7.

77. David Thomas, *Travels through the Western Country in the Summer of 1816* (Auburn, N.Y., 1819), 41.

CHAPTER 3: SUCKERS

1. The Rome Beauty is one example of a still popular variety that emerged from a sucker; Joan Morgan and Alison Richards, *The New Book of Apples* (London: Ebury, 2002), 262.

2. Creighton Lee Calhoun, Jr., *Old Southern Apples: History and Uses* (Blacksburg, Va.: McDonald & Woodward, 1995), 2–3.

3. Benjamin Rush, *Essays Literary, Moral, and Philosophica* (Schenectady, N.Y.: Union College Press, 1988), 130.

4. Thomas Ford, *A History of Illinois from Its Commencement as a State in 1818 to 1847* (Chicago, 1854), 68.

5. U.S. Federal Census, 1800, Longmeadow, Hampshire County, Mass. Although no death records for any of the children survive, one of the girls is absent from the 1800 census records, probably Patty, who would have turned ten that year.

6. Mortimer Elwyn Cooley, *The Cooley Genealogy: Descendants of Ensign Benjamin Cooley, an Early Settler of Springfield and Longmeadow, Massachusetts, and other Members of the Family in America* (Rutland, Vt.: Tuttle, 1941), 225–26, 240–42. Precisely when the Cooleys departed for Marietta is unknown, but four of them—Asahel, Jabez, Reuben, and Simeon—were granted title to donation tract lands north of Marietta in April 1797. Original conditions for securing a title to these lands included a five-year continuous settlement, but the rules appear to have been relaxed after the Indian threat to Marietta disappeared in 1795. Index to "Descriptions of land granted to settlers in the various allotments of the Ohio Company Donation Tract,"

Manuscripts and Documents of the Ohio Company of Associates, Marietta College Library, Marietta, Ohio, http://digicoll.marietta.edu/oca/document/S1OvD1I1/full/0004.html.

7. W. E. Peters, *Ohio Lands and their History* (Athens, Ohio: Messenger Printery, 1930), 180–85; "Ohio Company Donation Tract." For land bought and sold by Jabez, Simeon, and Asahel Cooley after 1799, see Deed Books, Athens County Recorder's Office, Athens, Ohio, 1:18, 73, 144, 318; 2:128, 212, 213; 3:77, 305, 306, 337, and Deed Books, Washington County Recorder's Office, Marietta, Ohio, 6:285; 9:10, 171; 10:74.

8. Robert Leslie Jones, *History of Agriculture in Ohio to 1880* (Kent, Ohio: Kent State University Press, 1983), 213–14, 217–18.

9. Deed records in Athens and Washington Counties record several purchases and sales by the Cooleys in the next two decades.

10. John Daniels ledger, May 14, 1798, and May 3, 1799, Warren County Historical Society, Warren, Penn.

11. For my understanding of the finer distinctions among financial instruments in the early republic, I am indebted to Pierre Gervais, lecturer, University of Paris St. Denis, and to Bruce Mann's *Republic of Debtors* (Cambridge, Mass.: Harvard University Press, 2002), 10–14.

12. W. E. Peters, *Ohio Lands and the History*, 3rd ed. (Athens, Ohio: W. E. Peters, 1930), 183–84.

13. The U. S. Federal Census, 1810, Fearing Township, Washington County, Ohio, lists Lucy Chapman living with two children, alongside Parley, living with his wife and one child. Oral tradition places the family further north near Dexter City in present-day Noble County, and Parley Chapman is buried there. Fearing Township is about fifteen miles south of the grave of Parley Chapman.

14. W. M. Glines, *Johnny Appleseed by One Who Knew Him* (Columbus, Ohio: F. J. Beer, 1922), 8.

15. H. Z. William, *History of Washington County, Ohio* (Cleveland, 1881), 544, 552.

16. U.S. Federal Census, 1820, Waterford Township, Washington County, Ohio; Gary S. Williams, *Johnny Appleseed in the Duck Creek Valley* (Caldwell, Ohio: Noble County Historical Society, 2004).

17. Elizabeth Cottle, "Chapman Families of Washington County," *The Tallow Light: Bulletin of the Washington County Historical Society* 18, no. 2 (1987–88): 69–70.

18. One preserved oral tradition of Chapman's visits to family is in a letter, Mrs. Burdette Harms to Robert Price, January 20, 1955, Robert Price Papers, Box 2, Folder 1, Ohio Historical Society, Columbus. Mrs. Harms claimed that Chapman frequently stopped at her great grandmother's house in Morgan County on his way to and from family visits, much to the chagrin of her great grandfather, who didn't much care for the wandering apple seed planter.

19. Jones, *History of Agriculture in Ohio*, 213–14, 217–18; Carlos Burr Dawes, "Johnny Appleseed in Marietta and the Muskingum Valley," *Tallow Light* 15, no. 1 (1984–85): 3–40, collects most of the Marietta traditions, most of them unsourced. Williams, *Johnny Appleseed in the Duck Creek Valley*, includes a few additional traditions from Noble County.

20. John A. Williams, "Our Cabin; or, Life in the Woods," *American Pioneer: A Monthly Periodical* 2, no. 10 (Oct. 1, 1843): 442–43, American Periodical Series Online (APS).

21. Ibid., 445, 451–55.

22. Ibid., 454–55.

23. J. A. Caldwell, *History of Belmont and Jefferson Counties* (1880), 191–92.

24. Ibid., 444, 587; James Morton Callahan, *Semi-Centennial History of West Virginia* (N.p.: Semi-Centennial Commission of West Virginia, 1913), 343.

25. Caldwell, *History of Belmont and Jefferson Counties*, 177.

26. William, "Our Cabin," 454.

27. Caldwell, *History of Belmont and Jefferson Counties*, 192. This story did not appear in print until 1880, and the writer of the county history asserted that Chapman's reluctance to plant trees locally was because he was offended by Nessly's grafted orchards nearby. This was likely an embellishment added by the author of the county history, as the words he puts in Chapman's mouth appear in other printed accounts that had been circulating for years before Caldwell's history was written.

28. Rosella Rice, "Johnny Appleseed" in *Arthur's Illustrated Home Magazine* 44 (December 1876): 634.

29. Douglas Hurt, *The Ohio Frontier* (Indianapolis: Indiana University Press, 1996), 178.

30. Alfred Byron Sears, *Thomas Worthington: Father of Ohio Statehood* (Columbus: Ohio State University Press, 1998), 19.

31. Ibid., 28–33.

32. David Mead Massie, *Nathaniel Massie: A Pioneer of Ohio* (Cincinnati, 1896), 62.

33. Ben F. Sager, *The Harrison Mansion* (Vincennes, Ind., 1928), 12–13.

34. Beverly W. Bond, Jr., "Civilization Comes to the Old Northwest," *Mississippi Valley Historical Review* 19, no. 1 (June 1932): 9–10. See also Ophia Smith, "Early Gardens and Orchards," *Bulletin of the Historical and Philosophical Society of Ohio* 7, no. 3 (1949): 67–89.

35. At least one of these trees grew to maturity, and decades later Mrs. Stadden was still sharing these apples, and her story, with her Licking Valley neighbors. Isaac Smucker, *Our Pioneers, Being Biographical Sketches of Capt. Elias Hughes* (Newark, Ohio, 1872), 15.

36. Cooley, *Cooley Genealogy*, 240.

37. Robert Price, *Johnny Appleseed: Man and Myth* (Urbana, Ohio: Urbana University, 2001), 284–85.

38. Horace Knapp, *A History of the Pioneer and Modern Times of Ashland County* (Philadelphia, 1863), 30–31. Smucker, *Our Pioneers*, gives brief biographies of many of the first families of Licking County and how they arrived.

39. Smucker, *Our Pioneers*, 15.

40. Price, *Johnny Appleseed*, 51–52, 72.

41. Richard W. Judd, *Untilled Garden* (New York: Cambridge University Press, 2009), 218–19, 244.

42. Rush, *Essays*, 213–15, 217–19, 222.

43. Timothy Dwight, *Travels in New England and New York* (Cambridge, Mass.: Harvard University Press, 1969), 1:xxxv, 7.

44. Malcolm J. Rohrbough, *Transappalachian Frontier*. Third Edition, (Bloomington: Indiana University Press, 2008), 207–8.

45. Charles M. L. Wiseman, *Pioneer Period and Pioneer People of Fairfield County, Ohio* (Columbus, Ohio: F. J. Heer, 1901), 106.

46. Rolland Lewis Whitson et al., *Centennial History of Grant County, Indiana, 1812 to 1912* (Chicago: Lewis, 1914), 1149.

47. Hervey Scott, *A Complete History of Fairfield County, Ohio* (Columbus, Ohio, 1877), 195.

48. Morgan and Richards, *New Book of Apples*, 262.

49. A. Banning Norton, *A History of Knox County, Ohio* (Columbus, Ohio, 1862), 50–51.

50. N. N. Hill, *History of Knox County, Ohio* (Mt. Vernon, Ohio, 1881), 224.

51. William Utter, *Frontier State: 1803–1825* (Columbus: Ohio Historical Society, 1942), 78; Charles C. Royce, *Indian Land Cessions in the United States* (1900; reprint, New York: Arno, 1971), 668–69.

52. Price, *Johnny Appleseed*, 59.

53. Norton, *History of Knox County*, 50, 51–53.

54. Hurt, *Ohio Frontier*, 175.

55. Joseph Walker to John Chapman, Sept. 14, 1809, Knox County Deeds, Knox County Recorder's Office, Mt. Vernon, Ohio, A:116.

56. N. N. Hill, *History of Coshocton County, Ohio* (Newark, Ohio, 1881), 567, 595; Hill, *History of Knox County*, 433.

57. T. S. Humrickhouse, "Johnny Appleseed," originally published in Hovey's *Magazine of Horticulture* in 1846, reprinted in the *Farmer's Cabinet and American Herd Book* 10 (May 15, 1846): 317, APS.

58. Firelands Historical Society, *The Firelands Pioneer* (Norwalk, Ohio, 1858–78), 1:15.

59. Hill, *History of Coshocton County*, 567, 595; Hill, *History of Knox County*, 489; Norton, *History of Knox County*, 128.

60. Reproduced in Robert C. Harris, *Johnny Appleseed Source Book* (Fort Wayne, Ind.: Allen County–Fort Wayne Historical Society, 1945), 11, 13.

61. Firelands Historical Society, *Firelands Pioneer*, 1:9–10.

62. Hill, *History of Knox County*, 224.

63. Price, *Johnny Appleseed*, 74.

64. Noel Kingsbury, *Hybrid: The History and Science of Plant Breeding* (Chicago: University of Chicago Press, 2009), 400.

65. George William Hill, *History of Ashland County, Ohio* (Cleveland, 1880), 411.

66. Knapp, *History*, 397, 535; E. Bonar McGlaughlin, *Pioneer Directory and Scrapbook of Richland County* (Mansfield, Ohio, 1887), 15; Firelands Historical Society, *Firelands Pioneer*, 1:9, 15.

67. Glines, *Johnny Appleseed*; R. I. Curtis, "John Chapman, alias 'Johnny Appleseed,'" *Ohio Pomological Society Transactions* (Columbus, Ohio, 1859), 68–69.

68. See Robert Owens, *Mr. Jefferson's Hammer* (Norman: University of Oklahoma Press, 2007), chap. 3–4, for a detailed account of Harrison's aggressive efforts to secure Indian land cessions.

69. William Cronon, *Changes in the Land: Indians, Colonists, and the Ecology of New England* (New York: Hill & Wang, 1983), 143, 153; Alfred W. Crosby, *Ecological Imperialism: The Biological Expansion of Europe, 900–1900* (New York: Cambridge University Press, 1986), 188–90.

70. Quoted in R. David Edmunds, *The Shawnee Prophet* (Omaha: University of Nebraska Press, 1983), 38.

71. Norton, *History of Knox County*, 516.

72. Reuben Gold Thwaites, ed., *Early Western Travels, 1748–1846* (New York: AMS Press, 1966), 4:227.

73. Edmunds, *Shawnee Prophet*, 115–16; Hurt, *Ohio Frontier*, 326–28.

74. Quoted in Alan Taylor, *The Civil War of 1812* (New York: Alfred A. Knopf, 2010), 157.

75. Ibid., 157–65.

76. Price, *Johnny Appleseed*, 83.

77. Ibid., 90–92.

78. W. W. Williams, *History of the Fire-Lands Compromising Huron and Erie Counties, Ohio* (Cleveland, 1879), 298–300; Firelands Historical Society, *Firelands Pioneer*, 1:9–15, 15:1080–81.

79. Price, *Johnny Appleseed*, 88.

80. Ibid., 94; W. D. Haley, "Johnny Appleseed—A Pioneer Hero," *Harper's Monthly Magazine* 63 (Nov. 1871): 832.

81. Price, *Johnny Appleseed*, 95.

82. Ibid., 98.

83. Ibid., 99.

84. "The Copus Battle Centennial," *Ohio Archeological and Historical Society Publications* 21 (1912): 379–95.

85. Knapp, *History*, 535.

CHAPTER 4: WALKING BAREFOOT TO JERUSALEM

1. Silas Chesebrough, "Journal of a Journey to the Westward," *American Historical Review* 37, no. 1 (Oct. 1931): 80.

2. Quoted in John Waller, *The Real Oliver Twist* (Cambridge, U.K.: Icon Books, 2005), 4–6. From Alexis de Tocqueville, *Journeys to England and Ireland* (New Haven, Conn.: Yale University Press, 1958), 107–8.

3. "Report of the Society for Printing, Publishing and Circulating the Writings of Emanuel Swedenborg," Manchester, England, Jan. 14, 1817, reprinted in Robert Price, *Johnny Appleseed: Man and Myth* (Urbana, Ohio: Urbana University, 2001), 120.

4. Colleen McDannell and Bernhard Lang, *Heaven: A History* (New Haven, Conn.: Yale University Press, 1988), 181–84.

5. Ibid., 189.

6. Ibid., 192.

7. Marguerite Beck Block, *The New Church in the New World* (New York: Swedenborg Publishing Association, 1984), 62.

8. Ibid., 64–65.

9. Ibid., 62–63, 73–75.

10. Ibid., 75.

11. "Biography No. X," *The New Church Independent and Monthly Review* 22, no. 3 (March 1874): 135–36; Block, *New Church*, 112.

12. Carl Theophilus Odhner, *Annals of the New Church* (Bryn Athyn, Pa.: Academy of the New Church, 1904), 451.

13. Price, *Johnny Appleseed*, 125–27.

14. Block, *New Church*, 114.

15. "The Lincoln Life-Mask and How It Was Made," *Century Illustrated Magazine* 23, no. 2 (Dec. 1881): 227, American Periodical Series Online (APS).

16. Nathan O. Hatch, *The Democratization of American Christianity* (New Haven, Conn.: Yale University Press, 1989), 5–9.

17. Baptist preacher David Barrow in 1795, quoted in Stephen Aron, *How the West Was Lost* (Baltimore: Johns Hopkins University Press, 1996), 172.

18. Price, *Johnny Appleseed*, 120.

19. Block, *New Church*, 172; *New Jerusalem Magazine*, new series, 6 (1881): 440; William Schlatter to John Chapman, Mar. 20, 1820, reprinted in Price, *Johnny Appleseed*, 126–27.

20. Block, *New Church*, 84.

21. Ibid., 69, 79; *History of the Philadelphia Bible Christian Church* (Philadelphia: J. B. Lippincott, 1922) 21–29; Karen Iacobbo and Michael Iacobbo, *Vegetarian America: A History* (Westport, Conn.: Praeger, 2004), 10–14; Colin Spencer, *Vegetarianism: A History* (New York: Da Capo, 2004), 243–44, "Society of Bible Christians in England," *Graham Journal of Health and Longevity* 3 (Aug. 31, 1839): 18, APS.

22. Block, *New Church*, 117–21.

23. Arthur Bestor, *Backwoods Utopias* (Philadelphia: University of Pennsylvania Press, 1970), 210–13; Block, *New Church*, 119–20.

24. Ophia Smith, "Beginnings of the New Jerusalem Church in Ohio," *Ohio State Archeological and Historical Quarterly* 66, no. 3 (1957): 255–56.

25. A. Banning Norton, *A History of Knox County, Ohio* (Columbus, Ohio, 1862), 133.

26. W. W. Williams, *History of the Fire-Lands, Comprising Huron and Erie Counties, Ohio* (Cleveland, 1879), 303–4.

27. M. L. Wilson, "John Chapman, Better Known as Jonny Appleseed," n.d. [ca. 1863], unidentified newspaper clipping in Muskingum County Pioneer Society Scrapbook, Muskingum University Archives, New Concord, Ohio.

28. Henry Howe, *Historical Collections of Ohio* (Cincinnati, 1847), 1:260.

29. McDannell and Lang, *Heaven*, 193.

30. Reprinted in Price, *Johnny Appleseed*, 13–14.

31. Rosella Rice, quoted in Henry A. Pershing, *Johnny Appleseed and His Time* (Strasbourg, Va.: Shenandoah Publishing House, 1930), 153–54.

32. E. Bonar McGlaughlin, untitled, undated clipping from the *Ohio Liberal*, in Roeliff Brinkerhoff scrapbook, Brinkerhoff Papers, Ohio Historical Society, Columbus, 22–25; Howe, *Historical Collections*, 2:486.

33. John W. Dawson, "Johnny Appleseed," *Fort Wayne Sentinel*, Oct. 21 and 23, 1871.

34. Obituary, *Fort Wayne Sentinel*, Mar. 22, 1845, reprinted in Robert C. Harris, *Johnny Appleseed Source Book* (Fort Wayne, Ind.: Allen County–Fort Wayne Historical Society, 1945), 15.

35. Quoted in McDannell and Lang, *Heaven*, 194.

36. Norton, *History of Knox County*, 130–31.

37. "Report of the Society," n.p.

38. Obituary, *Fort Wayne Sentinel*.

39. Fort Franklin stores ledger, Sept. 3, 1796, 35, Crawford County Historical Society, Meadville, Penn.

40. Horace Knapp, *History of Pioneer and Modern Times of Ashland County, Ohio* (Philadelphia, 1863), 37–38.

41. Ibid., 36.

42. Ibid., 38.

43. Norton, *History of Knox County*, 130.

44. Salathiel Coffinbury, letter to the *Mansfield Shield and Banner*, 1871, reprinted in Harris, *Johnny Appleseed Source Book*, 15.

45. Rosella Rice, quoted in Knapp, *History*, 32.

46. Ibid.

47. Iven Richey, quoted in Harris, *Johnny Appleseed Source Book*, 16.

48. Dawson, "Johnny Appleseed," 71.

49. "Estate of John Chapman," typewritten transcript by Wendell Paddock, Feb. 20, 1935, Ohio Historical Society, Columbus.

50. Howe, *Historical Collections*, 2:432.

51. R. I. Curtis, "John Chapman, alias 'Johnny Appleseed,'" *Ohio Pomological Society Transactions* (Columbus, Ohio, 1859), 69.

52. Coffinbury, letter.

53. Howe, *Historical Collections*, 2:432; Norton, *History of Knox County*, 129.

54. Knapp, *History*, 37.

55. Lydia Maria Child, *The American Frugal Housewife* (Boston, 1833), 8.

56. Curtis, "John Chapman," 69; Fort Franklin stores ledger, Sept. 3, 1796, 35; *Fire Lands Pioneer* 1:9, 11–12, 15; Williams, *History of the Fire-Lands*, 298–300.

57. Joseph Metcalfe, *Out of the Clouds, into the Light* (Philadelphia, 1872), 33.

58. Iacobbo and Iacobbo, *Vegetarian America*, 56; Colin Spencer, *Vegetarianism: A History* (New York: Four Walls, Eight Windows, 2002), 245, 256.

59. Alexis de Tocqueville, *Democracy in America: Abridged Edition* (New York: Harper Perennial, 2007), 300–301.

60. John Wigger, "Ohio Gospel: Methodism in Early Ohio," in *Center of a Great Empire*, ed. Andrew Cayton and Stuart Hobbs (Athens: Ohio University Press, 2005), 72–73.

61. Johnson and Wilentz, *The Kingdom of Matthias* (New York: Oxford University Press, 1995), 28, 31–33.

62. W. D Haley, "Johnny Appleseed—A Pioneer Hero," *Harper's Monthly Magazine*, November 1871, 835. Another version of the story appears in Howe, *Historical Collections*, 2:432.

CHAPTER 5: TO SERVE GOD OR MAMMON?

1. R. Douglas Hurt, *The Ohio Frontier: Crucible of the Old Northwest, 1720–1830* (Bloomington: Indiana University Press, 1996), 241; Robert Leslie Jones, *History of Agriculture in Ohio to 1880* (Kent, Ohio: Kent State University Press, 1983), 35.

2. Stewart Davenport, *Friends of the Unrighteous Mammon: Northern Christians and Market Capitalism, 1815–1860* (Chicago: University of Chicago Press, 2008), 1.

3. Matt. 6:24 and Luke 16:13, AV.

4. Adam Smith, *An Inquiry into the Wealth of Nations* (Hartford, Conn., 1804), 1:19.

5. For a brief overview of the history of early state banks in Ohio, see William T. Utter, *The Frontier State: 1803–1825* (Columbus: Ohio Historical Society, 1942), 266–73.

6. Reprinted in Robert Price, *Johnny Appleseed: Man and Myth* (Urbana, Ohio: Urbana University, 2001), 151.

7. John Daniels ledger, Warren County Historical Society, Warren, Penn.

8. A. Banning Norton, *A History of Knox County, Ohio* (Columbus, Ohio, 1862), 172–73, 183; Utter, *Frontier State*, 274–75.

9. A complete list of John Chapman's landholdings can be found in Price, "Appendix C: John Chapman's Land Holdings," in *Johnny Appleseed*, 273–76; Auditor of State, "School Land Sale Ledger, 1814–1818," 196, Ohio Historical Society, Columbus.

10. At fifty cents a day, John could earn ten dollars with twenty days' work, but if the family that hired him provided him with room and board, it would take longer. His seedling trees ideally sold for "a fippenny bit," meaning six and a quarter cents. Sometimes he took much less for his trees.

11. Hurt, *Ohio Frontier*, 175.

12. Richland County, Ohio, Deeds, 1:493, 521; Wayne County, Ohio, Deeds, 1:355; Auditor of State, "School Land Sale Ledger," B73.

13. Jones, *History of Agriculture in Ohio*, 27–28.

14. Norton, *History of Knox County*, 182.

15. Ibid., 184–89.

16. Silas Chesebrough, "Journal of a Journey to the Westward," *American Historical Review* 37, no. 1 (Oct. 1931): 80.

17. Ibid., 82–83.

18. Ibid., 83.

19. Ibid., 84.

20. Norton, *History of Knox County*, 183.

21. Ibid., 184–89.

22. Richland County Deeds, 1:482.

23. Ibid., 1:482, 523, 614; Wayne County Deeds, 4:396.

24. See John Matteson, *Eden's Outcasts: The Story of Louisa May Alcott and Her Father* (New York: W. W. Norton, 2007), 22–26, and Odell Shepard, *Pedlar's Progress: The Life of Bronson Alcott* (Boston: Little, Brown, 1937), 47–74.

25. For a brief explanation of economic values of most Americans in the seventeenth and eighteenth century, see John Lauritz Larson, *The Market Revolution in America: Liberty, Ambition, and the Eclipse of the Common Good* (New York: Cambridge University Press, 2010), 3–5. Also see Christopher Clark, *The Roots of Rural Capitalism: Western Massachusetts, 1780–1860* (Ithaca, N.Y.: Cornell University Press, 1990), chap. 2.

26. Quoted in Priscilla Carrington Kline, "New Light on the Yankee Peddler," *New England Quarterly* 12, no. 1 (Mar. 1939): 91.

27. Letitia Coe Tannehill, *History of John and Rachel Tannehill and Their Descendants* (1903; reprint, Cincinnati: Fitting, 2004), 8.

28. Matteson, *Eden's Outcasts*, 24.

29. Ibid., 26.

30. A. B. Alcott to William Andrus Alcott, March 1823, in A. Bronson Alcott, *New Connecticut* (Boston, 1887), 226–27.

31. Shepard, *Pedlar's Progress*, 130.

32. Tannehill, *History of John and Rachel Tannehill*, 8.

33. David Jaffee, "Peddlers of Progress, and the Transformation of the Rural North," *Journal of American History* 78, no. 2 (Sept. 1991): 514, 531.

34. Ibid., 528.

35. Ibid., 511–513.

36. Rosella Rice, "Johnny Appleseed," printed in James M'Gaw, *Philip Seymour* (Mansfield, Ohio, 1883), 430.

37. Biographer Robert Price places the Broom family cabin just outside of Perrysville. If Chapman did not keep the land near Mansfield for his sister Percis and her family, the importance of keeping that land remains a mystery.

38. Horace Knapp, *History of Pioneer and Modern Times of Ashland County, Ohio* (Philadelphia, 1863), 33.

39. Wayne County Deeds, 4:396, 397 and 32:46.

40. Knapp, *History*, 29–30.

41. Daniel Walker Howe, *What Hath God Wrought* (New York: Oxford University Press, 2007), 244. For a detailed discussion of the meaning of improvement to eastern agricultural reformers, see Steven Stoll, *Larding the Lean Earth: Soil and Society in Nineteenth-Century America* (New York: Hill & Wang, 2002), 19–30.

42. An excellent account of the Erie Canal story is Carol Sheriff, *The Artificial River: The Erie Canal and the Paradox of Progress, 1817–1862* (New York: Hill & Wang, 1996).

43. Francis P. Weisenburger, *The Passing of the Frontier: 1825–1850* (Columbus: Ohio Historical Society, 1941), 90, 94–97.

44. Albert Lowther Demaree, *The American Agricultural Press, 1819–1860* (Philadelphia: Porcupine Press, 1974), 36, 337.

45. Jones, *Agriculture in Ohio*, 400–401.

46. Ibid., 281–83.

47. Ibid.

48. W. D. Haley, "Johnny Appleseed—A Pioneer Hero," *Harper's Monthly Magazine*, November 1871, 834.

49. J. A. Caldwell, *History of Belmont and Jefferson Counties* (Wheeling, W.V., 1880), 192.

50. Knapp, *History*, 29.

51. John Henris, "Apples Abound: Farmers, Orchards, and the Cultural Landscapes of Agrarian Reform, 1820–1860" (Diss., University of Akron, 2009), 175–80.

52. Peter Cartwright, *Autobiography of Peter Cartwright* (New York: Abingdon, 1956), 141.

53. W. J. Rorabaugh, *The Alcoholic Republic: An American Tradition* (New York: Oxford University Press, 1979), 111.

54. See Benjamin Rush's "Moral Thermometer" in Rorabaugh, *Alcoholic Republic*, 44–45.

55. "Cider," *New York Observer and Chronicle*, Sept. 26, 1840, 156; quoted in Henris, "Apples Abound," 182.

56. "A Reformed Drunkard Ruined by Cider," *Zion's Herald* 6, no. 22 (June 3, 1835), American Periodical Series Online (APS).

57. "What Shall I Do with My Apples? BURN THEM," *Religious Intelligencer* 12, no. 19 (Oct. 6, 1827): 299, APS.

58. "Temperance," *Western Recorder* 7 (Dec. 28, 1836): 208, APS.

59. "Cider," *Western Recorder* 7 (Aug. 31, 1830): 138, APS.

60. "Another Lecture on Cider Drinking," *Western Christian Advocate* 3, no. 27 (Oct. 28, 1836): 107, APS.

61. "The Value of Apples for Stock," *Farmer & Gardener, and Livestock Breeder and Manager* 4, no. 19 (Sept. 5, 1837): 148, APS; *Western Christian Advocate, 3, no. 28* (Nov. 4, 1836): 110, APS.

62. "The Anecdotiad," *Masonic Mirror: Science, Literature, and Miscellany* 1, no. 2 (July 11, 1829): 13, APS.

63. "The Apple as Food," *Water-Cure Journal* 6, no. 1 (July 1, 1848): 28, APS; "Care of Apple Trees," *New Genesee Farmer and Gardener's Journal* 3, no. 7 (July 1842): 99, APS.

64. Henry David Thoreau and Bradley P. Dean, *Wild Fruits : Thoreau's Rediscovered Last Manuscript* (New York: W. W. Norton, 2000), 92.

65. "Apple as Food," 28; "Care of Apple Trees," 99.

66. *Fort Wayne Sentinel,* Oct. 21, 23, 1871.

67. Reprinted in H. Kenneth Dirlam, *John Chapman: By Occupation a Gatherer and Planter of Apple Seeds* (Mansfield, Ohio: Richland County Historical Society, 1956), 26.

68. N. N. Hill, *History of Coshocton County, Ohio* (Newark, Ohio, 1881), 293.

69. "Astonishing Profits from Ten Apple Trees," *National Register* 10, no. 2 (July 8, 1820): 25, APS.

70. Noel Kingsbury, *Hybrid: The History and Science of Plant Breeding* (Chicago: University of Chicago Press, 2009), 81–82.

71. Ibid., 129–31.

72. Ibid., 82.

73. "Johnny Appleseed," *Ohio Cultivator* 2, no. 17 (Sept. 1, 1846): 136, APS.

74. Hill, *History of Coshocton County,* 293.

75. Mercer County, Ohio, Deeds, B:14, 15, and C:177, 188–89; *History of Van Wert and Mercer Counties,* (Wapokaneta, Ohio, 1882), 121–22.

76. *History of Van Wert and Mercer Counties,* 121–22.

77. Dirlam, *John Chapman,* 236.

78. William Ellery Jones, ed., *Johnny Appleseed: A Voice in the Wilderness* (West Chester, Pa.: Chrysalis Books, 2000), 107–8. Jones claims Chapman knew Solomon Johnacake and Billy Dowdee when they lived at Greentown but does not provide a source for this information.

79. Mortimer Cooley, *The Cooley Genealogy: The Descendants of Ensign Benjamin Cooley, an Early Settler of Springfield and Longmeadow, Massachusetts* (Evansville, Ind.: Whipporwill, 1991), 243.

80. S. H. Mitchel, *The Indian Chief, Journeycake* (Philadelphia, 1895), 22–23; C. A. Weslager, *The Delaware Indians: A History* (New Brunswick, N.J.: Rutgers University Press, 1989), 360–61.

81. Wayne County Deeds, 4:397. According to Johnny Appleseed researcher William Ellery Jones, the purchaser of these lands, John Pile, was a New Churchman. If that is the case, it might explain John's generous offering price; Jones, *Johnny Appleseed,* 108–9.

82. Orestes Brownson, *Babylon Is Falling: A Discourse Preached in the Masonic Temple* (Boston, 1837), 4; Larson, *Market Revolution in America,* 93.

83. Matteson, *Eden's Outcasts,* 3–4.

84. Ibid., 156–59.

85. Hancock County, Ohio, Deeds, 3:118.

86. Entered at the Fort Wayne Land Office, May 16, 1838; Deed Register, State Archives, Indiana State Library, Indianapolis.

87. Rosella Rice, "Johnny Appleseed," *Arthur's Illustrated Home Magazine* 12 (1876): 636.

88. "The Estate Papers of John Chapman," transcript, Ohio Historical Society Archives, Columbus.

89. O. Morrow and F. W. Bashore, *Historical Atlas of Paulding County, Ohio* (Madison, Wis., 1892), 17. John Dawson's account, which first appeared in the *Fort Wayne Sentinel*, Oct. 21 and 23, 1871, is reprinted in Robert C. Harris, *Johnny Appleseed Source Book* (Fort Wayne, Ind.: Allen County–Fort Wayne Historical Society, 1945), 6–10.

90. Howe, *What Hath God Wrought*, 574.

91. Ben Sager, *The Harrison Mansion* (Vincennes, Ind., Frances Vigo Chapter, Daughters of the American Revolution, 1928), 12.

92. Kline, "New Light," 97.

93. "Hard Cider," *New York Evangelist* 11 (June 6, 1840): 23, APS.

94. "Hard Cider War," reprinted in *The Liberator* 10 (Aug. 21, 1840): 34, APS.

95. William Cobbett, *The American Gardener* (London, 1821), no. 33; William Cooper Howells, *Recollections of Life in Ohio* (Cincinnati, 1895), 60.

96. Howells, *Recollections*, 76.

97. Austin C. Burdick, "The Two Orchards," *Gleason's Pictorial Drawing-Room Companion* 5 (Aug. 27, 1853): 9, APS.

98. "Stealing Fruit," *Genesee Farmer* 2 (July 21, 1832): 29, APS.

CHAPTER 6: YANKEE SAINT AND THE RED DELICIOUS

1. "Estate of John Chapman," typewritten transcript by Wendell Paddock, Feb. 20, 1935, 10, 15, Ohio Historical Society, Columbus.

2. O. Morrow and F. W. Bashore, *Historical Atlas of Paulding County, Ohio* (Madison, Wis., 1892), 17.

3. W. M. Glines, *Johnny Appleseed by One Who Knew Him* (Columbus, Ohio: F. J. Heer, 1922), 11.

4. Data from U.S. Historical Census Browser, 1830 and 1840, University of Virginia Library, Charlottesville, http://mapserver.lib.virginia.edu/.

5. Francis P. Weisenburger, *The Passing of the Frontier: 1825–1850* (Columbus: Ohio Historical Society, 1941), 108–10.

6. Data from U.S. Census, 1840, Muskingum County, Ohio, U.S. Historical Census Browser, University of Virginia Library, Charlottesville, http://mapserver.lib.virginia.edu/.

7. Norris F. Schneider, *The Muskingum River: A History and Guide* (Columbus: Ohio Historical Society, 1968), 21, 24.

8. U.S. Federal Census, 1810, Washington County, Ohio; Gary S. Williams, *Johnny Appleseed in the Duck Creek Valley* (Caldwell, Ohio: Noble County Historical Society, 2004).

9. Elizabeth Cottle, "Chapman Families of Washington County," in *The Tallow Light: Bulletin of the Washington County Historical Society* 18, no. 2, (1987–88): 69–70.

10. Glines, *Johnny Appleseed*, 9–10.

11. Ibid., 11.

12. *Fort Wayne Sentinel,* March 22, 1845, reprinted in Robert C. Harris, *Johnny Appleseed Source Book* (Fort Wayne, Ind.: Allen County–Fort Wayne Historical Society, 1945), 18–19.

13. Harris, *Johnny Appleseed Source Book,* 10.

14. Dawson, "Johnny Appleseed," in Harris, *Johnny Appleseed Source Book,* 35.

15. Fort Wayne Sentinel, March 22, 1845, reprinted in Harris, *Johnny Appleseed Source Book,* 18–19.

16. J. S. Schenk, *History of Warren County* (Syracuse, N.Y., 1887) 153–54.

17. "Estate of John Chapman."

18. Bruce Mann, *Republic of Debtors* (Cambridge, Mass.: Harvard University Press, 2002), 10–14.

19. "Law for the Protection of Gardens, &c.," *Ohio Cultivator* 2, no. 6 (Mar. 15, 1846): 47, American Periodical Series Online (APS).

20. Robert Leslie Jones, *History of Agriculture in Ohio to 1880* (Kent, Ohio: Kent State University Press, 1983), 290.

21. "Johnny Appleseed," *Ohio Cultivator* 2, no. 17 (Sept. 1, 1846): 36, APS.

22. Henry Howe, *Historical Collections of Ohio* (Cincinnati, 1847), 431–32.

23. "Seed-Sowing," *German Reformed Messenger* 25, no. 51 (Aug. 17, 1859): 4, APS.

24. A brief biography and some of her writings can be found at "Rosella Rice—Her Stories," http://rosellarice.com/index_files/Page552.htm.

25. James M'Gaw, *Philip Seymour, or Pioneer Life in Richland County* (Mansfield, Ohio, 1883), 432.

26. A brief biography and an extended description of Haley's troubles with pro-slavery settlers in Alton, Illinois, is provided in Lewis Perry and Matthew C. Sherman, "What Disturbed the Unitarian Church in This Very City? Alton, the Slavery Conflict, and Western Unitarianism," *Civil War History* 54, no. 1 (2008): 18–34.

27. W. D. Haley's role in the formation of local Granges is documented in correspondence between Haley and Oliver Kelley, the moving force behind the Grange. Kelley and Haley may have first met each other when both were working in Washington, D.C., right after the Civil War. Oliver Hudson Kelley, *Origin and Progress of the Order of the Patrons of Husbandry in the United States* (Philadelphia, 1875), 223–24, 232–36, 248.

28. *Digest of the Laws and Enactments of the National Grange* (Philadelphia, 1882), 7.

29. D. Sven Nordin's *Rich Harvest: A History of the Grange, 1867–1900* (Jackson: University Press of Mississippi, 1974), 1–12, provides an overview of the Grange's origins and the role pomologist William Saunder played in its formation. Apple symbolism was embedded in the rituals of the Grange, with one of the thirteen officers of each chapter being dubbed "Pomona."

30. M'Gaw, *Philip Seymour,* 432.

31. Ibid., 431.

32. W. D. Haley, "Johnny Appleseed—A Pioneer Hero" *Harper's Monthly Magazine,* Nov. 1871, 83.

33. Lydia Maria Child, *The American Frugal Housewife* (Boston, 1833), 7.

34. An excellent biography of Child is Carolyn L. Karcher, *The First Woman in the Republic: A Cultural Biography of Lydia Maria Child* (Durham, N.C.: Duke University Press, 1994).

35. Lydia Maria Child, "Appleseed John," *St. Nicholas Magazine* 7, no. 8 (June 1880): 604, APS.

36. James M'Gaw, *Philip Seymour* (Mansfield, Ohio, 1857), 18.
37. Haley, "Johnny Appleseed," 832.
38. A. Banning Norton, *A History of Knox County, Ohio* (Columbus, Ohio, 1862), 133–34.
39. R. I. Curtis, "John Chapman, alias 'Johnny Appleseed,'" *Ohio Pomological Society Transactions* (Columbus, Ohio, 1859), 68–69; Glines, *Johnny Appleseed,* 10–11; Miscellaneous notes, E. B. Sturges Collection, Ohio Historical Society, Columbus.
40. "Seed-Sowing," 4.
41. Susan A. Dolan, *Fruitful Legacy: A Historic Context for Orchards in the United States* (Washington, D.C.: U.S. Government Printing Office, 2009), 32, 35.
42. A. J. Downing, *The Fruits and Fruit Trees of America* (New York, 1845).
43. Thomas A. Lyson, *Civic Agriculture: Reconnecting Farm, Food, and Community* (Medford, Mass.: Tufts University Press, 2004), 32, 35.
44. Dolan, *Fruitful Legacy,* 66–68.
45. Patricia Ortleib and Peter Economy, *Creating an Orange Utopia* (Westchester, Pa.: Swedenborg Foundation Publishers, 2011).
46. Dolan, *Fruitful Legacy,* 70–74.
47. Ibid., 63–68.
48. Vachel Lindsay, *The Litany of Washington Street* (New York: Macmillan, 1929), 75. *The litany of Washington street* (New York: The Macmillan Company, 1929
49. Ann Massa, *Vachel Lindsay: Fieldworker for the American Dream* (Bloomington: Indiana University Press, 1970); Edgar Lee Masters, *Vachel Lindsay: A Poet in America* (New York: Charles Scribner's Sons, 1935).
50. Vachel Lindsay, *Collected Poems*, rev. and illustrated ed. (New York: Macmillan, 1925), 89.
51. Ibid., 85.
52. Rosemary Benét and Stephen Vincent Benét, *A Book of Americans* (New York: Farrar & Rinehart, 1933), 47–48.
53. Lewis Mumford, *Faith for Living* (New York: Harcourt, Brace, 1940), 274–75.
54. Howard Fast, *The Tall Hunter* (New York: Harper, 1942), book jacket.
55. Howard Fast, *Being Red* (Boston: Houghton-Mifflin, 1990).
56. Walt Disney Studios, *Melody Time* (1948; Burbank, Calif.: Disney DVD, 2004), DVD.
57. Louis Bromfield, "The World Needs More Johnny Appleseeds," *Cincinnati Enquirer,* July 31, 1955, photocopy from the Murdock Collection, Johnny Appleseed Society Archives, Urbana University, Urbana, Ohio.
58. David Cook, "America's Barefoot Entrepreneur," *American Fruit Grower* 1 (1981): 32, 42–43.
59. Dolan, *Fruitful Legacy,* 118–24.
60. Aaron B. Wildavsky, *But Is It True?: A Citizen's Guide to Environmental Health and Safety Issues* (Cambridge, Mass.: Harvard University Press, 1995), 201–22; Jane Gregory and Steve Miller, *Science in Public: Communication, Culture, and Credibility* (New York: Plenum Trade, 1998), 168–73.
61. Dolan, *Fruitful Legacy,* 129.
62. Wendy Gordon, "The True Alar Story IV," May 12, 2011, *Huffington Post,* www.huffingtonpost.com/wendy-gordon/food-chemicals-_b_859179.html.

63. Dominique Browning, "Alar in Applesauce and Mercury in Fish," June 23, 2011, *Mom's Clean Air Force*, www.momscleanairforce.org/2011/06/23/alar-in-applesauce-and-mercury-in-fish-the-connection-2/.

64. "Responsible Choice," Stemilt Growers LLC, www.stemilt.com/Farm_With_Us/ResponsibleChoice.cfm (accessed Mar. 5, 2012).

65. Elliott Negin, "The Alar 'Scare' Was for Real," *Columbia Journalism Review* 35, no. 3 (Sept.–Oct. 1996): 13–15.

66. Gordon, "True Alar Story IV"; Wildavsky, *But Is It True?*, 201–22; Gregory and Miller, *Science in Public*, 168–73.

67. Dolan, *Fruitful Legacy*, 130.

68. Ibid.

69. Lyson, *Civic Agriculture*, 33, 36.

70. *Broken Limbs: Apples, Agriculture, and the New American Farmer*, directed by Guy Evans (Oley, Pa.: Bullfrog Films, 2004), DVD; Dolan, *Fruitful Legacy*, 114–15.

71. "Legislative Priorities," U.S. Apple Association, www.usapple.org/government/legislative-priorities (accessed Aug. 1, 2011).

72. Tammi Flynn and Jenne Drury, *The "3 Apple-a-Day" Plan: Your Foundation for Permanent Fat Loss* (Wenatchee, Wash.: Get Fit, 2003).

73. Dolan, *Fruitful Legacy*, 142, 46.

74. Ibid.

75. "The Project," Johnny Appleseed Heritage Center Inc., www.jahci.org/project.html (accessed Mar. 5, 2012).

76. Robert Price, *Johnny Appleseed: Man and Myth* (Urbana, Ohio: Urbana University, 2001), 338–46. More information about the Johnny Appleseed Society and Museum can be found at www.urbana.edu/index.php/alumni_and_friends/appleseed_society/.

77. Michael Pollan, *Botany of Desire* (New York: Random House, 2001), 39, 21–22.

78. David Diamond, "Origins of Pioneer Apple Orchards in the American West: Random Seeding versus Artisan Horticulture," *Agricultural History Review* 84, no. 4 (Fall 2010): 424–25.

79. Howard Means, *Johnny Appleseed: The Man, the Myth, the American Story* (New York: Simon & Schuster, 2011), 274.

80. Robert Morgan, *Lions of the West: Heroes and Villains of the Westward Expansion* (Chapel Hill, N.C.: Algonquin, 2011), 115, 113; Dale L. Walker, "Of the Heroes Heading West, Some Were More Leonine than Others," *Dallas Morning News*, October 23, 2011.

81. I borrow the phrase "inherited a revolution" from Joyce Appleby, *Inheriting the Revolution: The First Generation of Americans* (Cambridge, Mass.: Belknap Press of Harvard University Press, 2000).

82. A brief history of Sholan Farms can be found at http://sholanfarms.com/history/ (accessed Mar. 5, 2012).

83. *Broken Limbs*.

84. Creighton Lee Calhoun, *Old Southern Apples: A Comprehensive History and Description of Varieties for Collectors, Growers, and Fruit Enthusiasts*, 2nd ed. (White River Junction, Vt.: Chelsea Green, 2010).

85. "The Applesource Story," Applesource, www.applesource.com/asstory.html (accessed Mar. 5, 2012).

86. "Apples Top New 'Dirty Dozen' List," Environmental Working Group, news release, June 13, 2011, www.ewg.org/release/ewgs-2011-shoppers-guide-helps-cut-consumer-pesticide-exposure.

87. Darrin Nordahl, *Public Produce: The New Urban Agriculture* (Washington, D.C.: Island, 2009); Lyson, *Civic Agriculture*.

88. Fallen Fruit, www.fallenfruit.org (accessed Mar. 5, 2012).

89. Portland Fruit Tree Project, http://portlandfruit.org (accessed Mar. 5, 2012).

90. City Fruit, www.cityfruit.org (accessed Mar. 5, 2012).

91. "TEDxBoston—Lisa Gross—Civic Fruit," July 13, 2011, www.youtube.com/watch?v=9ZEfvrgB78E.

Essay on Sources

John Chapman and his family left few traces on the historical record. But in vital records, federal and local censuses, land deeds, country store ledgers, IOUs, and estate papers, a sketch of John his life emerges. Some of these documents offer clear and direct evidence of John Chapman's whereabouts and activities in a specific time and place. Others, like the two imperfectly constructed bills of exchange for a hundred dollars he drew up in Franklin, Pennsylvania, in 1804 left me with unanswered questions, possibilities and probabilities but no certainties. Yet careful study of these disparate sources, scattered across four states, provide more than just information about John Chapman's comings and goings and his changing economic strategies. They also reveal much about the worlds he inhabited and how he fit into them. Ledger books from dry goods stores tells us not just what John Chapman bought but also who his neighbors were, and how they lived. The John Daniels ledger at the Warren County Historical Society, and the Holland Land Company ledgers in the collections of the Crawford County and Franklin County Historical Societies, all in Pennsylvania, were invaluable in this regard. Federal censuses illustrate the stages of settlement of communities in which Chapman resided. IOUs and land deeds offer a wealth of information not just about John Chapman's economic activities but also regarding his neighbors'.

Contemporaneous records not directly related to John Chapman were also invaluable. The Irvine-Newbold Papers at the Pennsylvania Historical Society provide a wealth of information about General William Irvine and his son Callender and their activities in northwestern Pennsylvania around the turn of the nineteenth century. In the letters Callender Irvine wrote to his speculator father from his primitive land claim on Brokenstraw Creek, Chapman's shadow steals across the page as one of the rough-looking men Callender feared were out to jump his claim. The memoir of Cornelius Van Horn, located in the Crawford County Historical Society, helped me puzzle out the choices and challenges young single men faced on the northwest Pennsylvania frontier. The published records of the legal disputes between speculators and squatters in northwest Pennsylvania, *The Report of the Case of the Commonwealth vs. Tench Coxe* (Philadelphia, 1803), include depositions from many settlers involved in the conflict. The frantic land acquisitions and sales of John Chapman's Cooley step-cousins in Ohio Company lands, recorded in the deed registers of Washington, Morgan, and Athens County, Ohio, reveal what it took to acquire land in the early years of settlement near the Muskingum, and why neither John nor his father were able to do so. Records of the Ohio Land Company in the Special Collections of the Marietta College Library were also useful in this regard.

Pioneer accounts and travel diaries from the places John Chapman passed through in Ohio describe the state of horticulture in specific times and places, which helped me understand the logic behind some of Chapman's early Ohio wanderings. Among the more useful of these were Fortescue Cuming, *Sketches of a Tour of the Western Country* (Pittsburgh, 1810); Silas Chesebrough, "Journal of a Journey to the Westward," *American Historical Review* 37 (Oct. 1931); and John A. Williams, "Our Cabin; or, Life in the Woods," *American Pioneer: A Monthly Periodical*, Oct. 1, 1843, American Periodical Series Online (APS).

The character and personality of John Chapman were preserved in oral traditions. Only those saved in the reports and correspondence of members of the New Church in England and America were recorded in print during Chapman's lifetime. The bulk of the oral tradition about John Chapman emerged in the quarter century after his death. These appeared in letters to horticultural journals and in the first state and county histories. Among the most important of these are the accounts that appeared in Henry Howe's *Historical Collections of Ohio* (Cincinnati, 1847), R. I. Curtis's letter in the *Ohio Pomological Society Transactions* (1859), A. Banning Norton's *History of Knox County* (Columbus, Ohio, 1862), and H. S. Knapp's *History of the Pioneer and Modern Times of Ashland County* (Philadelphia, 1863). Such sources, which contain mundane information as well as tall tales, have to be approached with great care. My approach has been to break these down into their component parts; to separate the varied aspects of Chapman's physical description, his behavior, and the peculiar stories told about him and to construct a sort of "genealogy" of the myth. By doing so, it became clearer where particular stories initially emerged and how these stories were often borrowed, adapted, and reused in subsequent printed accounts.

As published stories about John Chapman began circulating, they sometimes prompted rebuttals. When Chapman's story first reached a truly national audience in "Johnny Appleseed—A Pioneer Hero," *Harper's Monthly Magazine*, November 1871, a few responses followed. Catherine Stadden's dismissive account of John Chapman's alleged apple tree planting activities in Licking County, Ohio, printed in Isaac Smucker, *Our Pioneers*... (1872), and Fort Wayne, Indiana, resident John Dawson's refutation of the *Harper's* article are two important examples. While these authors obviously could have their own motivations for inventing, distorting, or exaggerating, they were nonetheless extremely helpful in evaluating the authenticity of specific stories and in countering the early efforts at hagiography. Also, accounts of Johnny Appleseed that appeared in county histories and popular magazines in the late nineteenth century, and across the twentieth and into the twenty-first, were valuable primary source of a different sort—in tracing the evolution of the Johnny Appleseed myth. Rosella Rice, who grew up in Perrysburg, Ohio, and knew John Chapman when she was a young girl, became a prolific writer of sentimental fiction in adulthood and produced several accounts of John Chapman, based, to varying degrees, on her childhood memories. Her version of Johnny Appleseed's story, which she continually supplemented and revised, appeared in books and popular magazines in the late nineteenth century. Nearly every Ohio county produced a county history in the late nineteenth or early twentieth century. Johnny Appleseed stories appeared in dozens of them. While many borrowed from other volumes, unique local Appleseed stories regularly appeared.

Finally, this work is also informed by a large and varied body of primary sources on apples and apple culture in North America, from the time of the earliest English settlements to the eve of the American Civil War. Much of that is invisible in the narrative, but it undergirds my understanding of apples and their changing uses and meaning. I depended on a wide range of published travel accounts from early America for knowledge of horticultural practices and read deeply in the many English and American books and manuals on horticulture available to agricultural improvers in the late eighteenth and early nineteenth century. The most important American books on horticulture in these years include William Coxe, *A View of the Cultivation of Fruit Trees* (Philadelphia, 1817), William Kenrick, *The New American Orchardist* (Boston, 1835), and Andrew Jackson Downing, *The Fruits and Fruit Trees of America* (New York, 1847). While many of these titles are now accessible via Google Books, I was delighted to have the opportunity to peruse original copies of these and many other titles in the collections of the William L. Clements Library in Ann Arbor, Michigan.

Agricultural journals from the early nineteenth century, including the *Genessee Farmer,* the *American Farmer,* the *Western Farmer and Gardener,* and the *Ohio Cultivator,* were also invaluable sources. Many of these are included in the American Periodical Series online database. At the Filson Historical Society in Louisville, Kentucky, the *Franklin Farmer,* the *Kentucky Farmer,* and the *Farmer's Manual* were useful sources.

The personal papers of wealthier farmers, describing in some detail their horticultural activities, were also valuable. The diary of Joseph Hornsby, an early planter in Shelby, Kentucky, in the Filson Historical Society's collections, was immensely helpful for the information it conveyed about the significant expense and labor required to establish grafted orchards in the West. The papers of Laommi Baldwin in the Clements Library in Ann Arbor were also useful. Baldwin is credited with the promotion, if not discovery, of the Baldwin Apple, an American cultivar that emerged as one of the most popular varieties by the middle of the nineteenth century.

The Urbana University Library in Urbana, Ohio, contains perhaps the most important collection of American materials on Swedenborg and the New Church in America. The archives of the Johnny Appleseed Society Museum, also located on the Urbana campus, house a rich collection of representations of Johnny Appleseed in popular culture. The collections of the Ohio Historical Society in Columbus contain materials on the life of John Chapman, including the papers of Robert Price, John Chapman's first biographer. Price spent seventeen years researching and writing his 1954 book on John Chapman and gathered most of the discovered primary sources pertaining to his life.

SECONDARY SOURCES

Crafting the story of John Chapman—a man who left a faint trace on the historical record but a large one in myth—and setting it in a deep historical context presented many unique challenges, so early on in my research I sought out examples of books by authors who faced similar challenges. While I cannot say with any confidence that I emulated their approach or learned specific tricks from them, there are several works I deeply admire that shaped my thinking about Johnny Appleseed. Among these are Michael Allen's *Western Rivermen, 1763–1861: Ohio and Mississippi Boatmen and the Myth of the Alligator Horse* (Lousiana State University Press, 1990), which ably recovers the story of the legendary Mike Fink. I also benefited from other "microhistories" published in the last two decades, which served as models. Among these were Paul Johnson and Sean Wilentz, *The Kingdom of Matthias: A Story of Sex and Salvation in Nineteenth-Century America* (Oxford University Press, 1994); John Demos, *The Unredeemed Captive: A Family Story from Early America* (Vintage Books, 1994); Alfred Young, *The Shoemaker and the Tea Party: Memory and the American Revolution* (Beacon, 1999); and Paul Johnson, *Sam Patch, Famous Jumper* (Hill &Wang, 2003).

Part of the appeal of writing a microhistory was knowing that it would require me to immerse myself in a broad body of scholarship. Understanding John Chapman's mission and how it was received by the communities he served required a knowledge of the changing uses of apples in early America. Noel Kingsbury's *Hybrid: The History and Science of Plant Breeding* (University of Chicago Press, 2009) offers a macrohistorical perspective on changing agricultural practices from premarket peasant societies to the present. Barrie E. Juniper and David J. Mabberly's *The Story of the Apple* (Timber Press, 2006) gives an accessible overview of the botany and history of the apple over time. U. P. Hedrick's *History of Horticulture in America to 1860* (Oxford University Press, 1950) provides a wealth of information on the apple's development in colonial and antebellum America. Regional studies such as Howard S. Russell's *Long Deep Furrow: Three Centuries of Farming in New England* (University Press of New England, 1982) and Robert Leslie Jones's *History of Agriculture in Ohio to 1880* (Kent State University Press, 1983) were also important. John Henris's excellent dissertation, "Apples Abound: Farmers, Orchards, and the Cultural Landscapes of Agrarian Reform, 1820–1860" (University of Akron, 2009), describes the shift from cider apples to winter apples in New England and Ohio. Susan A. Dolan, *Fruitful Legacy: A Historic Context of Orchards in the United States* (National Park Service, 2009), gives an overview of orcharding practices in the United State from colonial times to the present. Andrea Wulf, *Founding Gardeners: The Revolutionary Generation, Nature and the Shaping of the American Nation* (Alfred A. Knopf, 2011), offers important insights into the horticultural practices and the meanings of gardens and orchards to the nation's political and cultural elite in the early years of the republic. Finally, Philip Pauly's *Fruits and Plains: The Horticultural Transformation of America* (Harvard University Press, 2007) provides a history of horticultural science in the United States from the American Revolution to the present.

Dolan's *Fruitful Legacy* and Paul Conkin's *A Revolution Down on the Farm: Agriculture since 1929* (University Press of Kentucky 2009) helped me carry the story of the apple to the present day. For the movement against globalized factory farming, I relied on John E. Ikerd, *Crisis and Opportunity: Sustainability in American Agriculture* (Bison Books, 2008); Darrin Nordahl, *Public Produce* (Island Press, 2009); and the excellent documentary film *Broken Limbs: Apples, Agriculture, and the New American Farmer* (Bullfrog Films, 2004).

My understanding of the role European orchard agriculture played in the colonization of New England and the Midwest is deeply influenced by William Cronon's seminal work *Changes in the Land: Indians, Colonists, and the Ecology of New England* (Hill & Wang, 1983). Virginia Anderson's *Creatures of Empire: How Domestic Animals Transformed Early America* (Oxford University Press, 2004) offers a deeper evaluation of the impact of livestock in the early British North American colonies. My essay, "Apples on the Border: Orchards and the Contest for the Great Lakes," *Michigan Historical Review* 34, no. 1 (Spring 2008) examines the role European and Indian orchards played in the contest for that region. Finally, Brian Donohue's study of Concord, *The Great Meadow: Farmers and the Land in Colonial Concord*

(Yale University Press, 2004), is an excellent examination of agricultural transitions at the local level from Native American agriculture into the eighteenth century.

My portrait of Edward Chapman's world was informed by a wealth of secondary literature on early New England Society. What follows is just a sampling. Joseph A. Conforti, *Saints and Strangers: New England in British North America* (Johns Hopkins University Press, 2006), provides an overview of seventeenth- and eighteenth-century New England. David Hall, *A Reforming People: Puritanism and the Transformation of Public Life in New England* (Alfred A. Knopf, 2011), restores the reformist elements of the Puritan experiment, absent from many recent histories of the region, by reminding us of the leveling implications of the practice of partible inheritance. Virginia DeJohn Anderson, *New England's Generation: The Great Migration and the Formation of Society and Culture in the Seventeenth Century* (Cambridge University Press, 1992), illustrates the importance New Englanders placed on passing on a competency to their children. Daniel Vickers, *Farmers and Fishermen: Two Centuries of Work in Essex County, Massachusetts, 1630–1850* (University of North Carolina Press, 1994); Philip Greven, *Four Generations: Population, Land, and Family in Colonial Andover, Massachusetts* (Cornell University Press, 1972); and Kenneth A. Lockridge, *A New England Town: The First Hundred Years, Dedham, Massachusetts 1636–1736* (W. W. Norton, 1985), trace the challenges New England communities faced over time.

For comprehending Nathaniel Chapman's world, I am indebted to several works. Brian Donohue's *Great Meadow*, and Daniel Vickers's *Farmers and Fisherman*, both mentioned above, were invaluable for understanding transformations in farming, landownership, and labor that occurred in the eighteenth century. John Shy, *A People Numerous and Armed* (Oxford University Press, 1976), and Robert Gross, *The Minutemen and Their World* (Hill & Wang, 1976), helped set the context of Nathaniel Chapman's military service. Demos, *Unredeemed Captive*, offers insights into life in Longmeadow under the pastorate of Stephen Williams, and Leonard L. Richards, *Shays's Rebellion: The American Revolution's Final Battle* (University of Pennsylvania Press, 2002), provides an excellent background for understanding the social and economic turmoil that fostered that rebellion.

Richard White's *The Middle Ground: Indians, Empires, and Republics in the Great Lakes Region, 1650–1815* (Cambridge University Press, 1991) continues to influence scholarship on places where indigenous peoples and colonizers come into contact. Alan Taylor's *The Divided Ground: Indians, Settlers, and the Northern Borderland of the American Revolution* (Alfred A. Knopf, 2006) establishes the context for the story of white-Indian interaction in northwestern Pennsylvania at the end of the eighteenth century. Several books were critical in shaping my understanding of the Indian world John Chapman entered in these years: Matthew Dennis, *Seneca Possessed: Indians, Witchcraft, and Power in the Early American Republic* (University of Pennsylvania Press, 2010); David Swatzler, *A Friend among the Senecas: The Quaker Mission to Cornplanter's People* (Stackpole Books, 2000); Thomas S. Abler, *Cornplanter: Chief Warrior of the Allegany Senecas* (Syracuse University Press, 2007); and Anthony F. C. Wallace, *The Death and Rebirth of the Seneca* (Vintage, 1969).

On the contests that often pitted settler against settler and squatter against speculator for land in Pennsylvania, see, Patrick Griffin, *American Leviathan: Empire, Nation, and the Revolutionary Frontier* (Hill & Wang, 2007); Paul Moyer, *Wild Yankees: The Struggle for Independence along Pennsylvania's Revolutionary Frontier* (Cornell University Press, 2007); Robert

Arbuckle, *Pennsylvania Speculator and Patriot: The Entrepreneurial John Nicholson, 1757–1800* (Pennsylvania State University Press, 1975); and Paul D. Evans, *The Holland Land Company Purchase* (Augustus M. Kelley, 1924).

Ann Smart Martin's *Buying into a World of Goods: Early Consumers in Backcountry Virginia* (Johns Hopkins University Press, 2008) helped me interpret the many ledger books I encountered in northwestern Pennsylvania and Ohio. Stephen Mimh's *A Nation of Counterfeiters: Capitalists, Con Men, and the Making of the United States* (Harvard University Press, 2009) and Bruce Mann's *Republic of Debtors: Bankruptcy in the Age of American Independence* (Harvard University Press, 2009) aided me in understanding the mediums of exchange in early America. Scott Sandage's *Born Losers: A History of Failure in America* (Harvard University Press, 2005) is an excellent interpretation of the psychology of failure in the nineteenth century.

R. Douglas Hurt, *The Ohio Frontier: Crucible of the Old Northwest, 1720–1850* (Indiana University Press, 1998), and Malcolm J. Rohrbough, *Trans-Appalachian Frontier: People Societies and Institutions, 1775–1850*, 3rd ed. (Indiana University Press, 2008), are good starting points for understanding early white settlement in Ohio. John Mack Faragher, *Sugar Creek: Life on the Illinois Prairie* (Yale University Press, 1998), and Susan Sessions Rugh, *Our Common Country: Early Farming, Culture, and Community in the Nineteenth-Century Midwest* (Indiana University Press, 2001), while focusing on Illinois settlement, offer important insights into the formation of new communities on the Midwestern frontier.

I believe that changes in the ways ordinary Americans engaged in economic exchange profoundly shaped individual belief and behavior in the early decades of the nineteenth century. These changes are manifest in such simple things as the types of apples a farmer chose to grow. Many scholars today dismiss distinctions between "self-provisioning" and "market-oriented" farmers as overwrought, but the proliferation of journals promoting agricultural improvement in these years expose a real cultural divide among the nation's farmers. For an interpretation of the era presented by a "market revolution" skeptic, see Daniel Walker Howe, *What Hath God Wrought: The Transformation of America, 1815–1848* (Oxford University Press, 2007). For a contrasting view, see John Lauritz Larson, *The Market Revolution in America: Liberty, Ambition, and the Eclipse of the Common Good* (Cambridge University Press, 2010). Steven Stoll's *Larding the Lean Earth: Soil and Society in Nineteenth-Century America* (Hill & Wang, 2002), while focusing on the mindset of Eastern agricultural improvers, strikes a middle ground, suggesting that farmers' reactions to expanding markets represented a spectrum rather than two poles. Another work on the impact of the market on ordinary Americans that has influenced my perspective on the debate is Christopher Clark's *The Roots of Rural Capitalism: Western Massachusetts, 1780–1860* (Cornell University Press, 1990).

Since the 1990s, there has been much scholarship on the intersection between markets and morality in the early Republic. In *The Market Revolution: Jacksonian America, 1815–1846* (Oxford University Press, 1991), Charles Sellers argues that the American religious landscape in this era became divided into two broadly anti-market and pro-market faiths. Sellers's interpretation has faced a range of challenges in recent years, with most critics finding American church leaders had no trouble reconciling theology and capitalism's invisible hand. The majority of this scholarship was written prior to the financial collapse of 2008. It remains to be seen how revived contemporary concerns about the morality of the marketplace might influence subsequent scholarship. Essays in *God and Mammon: Protestants, Money, and the Market, 1790–1860*, ed. Mark A. Noll (Oxford University Press, 2001), and a special issue of *Early American Studies* 8, no. 3 (Fall 2010) on "Markets and Morality," ed. Cathy Matson, include several fresh perspectives. Stewart Davenport's excellent *Friends of the Unrighteous Mammon: Northern Christians and Market Capitalism, 1815–1860* (University of Chicago Press, 2008) focuses on ministers and other leading public Christians but offers useful categories for explaining a range of Protestant responses to the market. Finally, a collection of essays, *Thrift and Thriving in America: Capitalism and Moral Order from the Puritans to the Present*, ed. Joshua J. Yates and James Davison Hunter (Oxford University Press, 2011), challenged me to think more deeply about the meaning of Chapman's extreme commitment to frugality.

John Chapman matured in a world brimming with new religious choices. For my understanding of the implications of this development, I am indebted to Nathan O. Hatch, *The Democratization of American Christianity* (Yale University Press, 1989). For the spread of Swedenborgian ideas in America, I depended on Marguerite Block, *The New Church in the New World* (Swedenborg Publishing Association, 1984). Colleen McDannell and Bernhard Lang, *Heaven: A History* (Yale University Press, 1988), offer an important analysis of some of the innovations in Swedenborgian theology. David Paul Nord's *Faith in Reading: Religious Publishing and the Birth of Mass Media in America* (Oxford University Press, 2004) is an excellent study of the spread of print evangelism in nineteenth-century America. Johnson and Wilentz, *Kingdom of Matthias*, provide valuable insights into religious radicalism in the early decades of the nineteenth century and helped me understand movements that emphasized self-denial and primitivism as responses to the new materialism of the marketplace. John Matteson's *Eden's Outcasts: The Story of Louisa May Alcott and Her Father* (W. W. Norton, 2007) provides a window onto Bronson Alcott's own struggles for perfectionism in these years.

Allen, *Western Rivermen*; John Mack Faragher, *Daniel Boone* (Henry Holt, 1992); and Scott Reynolds Nelson, *Steel Drivin' Man: John Henry, the Untold Story of an American Legend* (Oxford University Press, 2006), provided models for explaining the evolution of the myth. John Conforti's *Imagining New England* (University of North Carolina Press, 2000) helped me understand the ways in which the Johnny Appleseed myth reflected the image of the Yankee across the nineteenth century.

Index

Adena (estate), 77, 82
Agawam, 7–8, 12
agricultural press, 141–42, 145, 167
agricultural reformers, 141–43, 146, 148, 149
agricultural societies, 140–42, 150, 161, 167
agriculture: European, 6, 9–11, 43, 44, 47–49; Native American, 7–8, 43, 45–47, 48–49
Alar, 182–85, 193
Alcott, Abba May, 136, 155
Alcott, Bronson, 123, 124, 135–36, 154–55
American Frugal Housewife (Child), 122, 171–72
American Revolution, 3, 20–27, 30, 31, 41, 54
applejack, 144
apples: campaign against theft of, 159–61, 167; and chemical treatment, 175–76, 182–86; and globalized markets, 185–87; Golden Delicious, 184–85; and national markets, 170, 175–76; Red Delicious, 4, 182–84, 189, 191, Rhode Island Greening, 9–10, 68; Rome Beauty, 82, 175; winter, 74, 148, 160
"Apple-Seed John" (Child), 172, 174
apple trees: dwarf-stock, 182, 184, 188; North American crab (*Malus coronia*), 5; Old World (*Malus pumila* or *Malus domestica*), 4–6. *See also* nurseries; orchards
apple trees, grafted: 4–6, 10–11; and John Chapman, 58–59, 149, 157; disadvantages of, 89, 148–49; and Eastern Ohio, 73–74; and Licking Valley, 77–79; and Marietta, 68–69; and market economy, 79–83, 142–43, 160–61; and top-grafting, 137, 148; and Virginia Military District, 76–78
apple trees, seedling ("wild apples"): 4, 5, 6, 10–11, 42; and John Chapman, 4, 59, 82–83, 85–87, 139, 142–43; and hogs, 40, 146, 148, 191; and the poor, 89, 137, 161, 191; and self-provisioning farmers, 6, 11, 88–89, 148–49; top-grafting, 74, 80, 82, 137, 148
Asbury, Francis, 124

bachelorhood, 60, 62–63, 103, 105, 116–17, 128, 190
Bailey, Francis, 106, 113, 139
Ballou, Hosea, 115
banks, 88, 100, 110, 129–31, 133–34, 154
barter, 31, 40, 59, 88–89, 110, 129
Benét, Rosemary and Stephen Vincent, 177–78
Bible Christians, 112, 123
"Big Bottom massacre," 68
bills of exchange, 70–71, 166
Black Fork, Ohio, 84, 90, 101–2, 132, 133, 153
Blaxton (Blackstone), William, 9–10
"book farmers," 142, 143, 176
Boone, Daniel, 2, 3, 49, 173
Bordeaux Mixture (hydrated copper sulfate), 175
Boston Tree Party, 193–94
Botany of Desire (Pollan), 188–89
Brodhead, Daniel, 41, 44
Broken Limbs (documentary), 192
Brokenstraw Creek, Pennsylvania, 37, 47, 48, 50–51, 54, 56
Bromfield, Lewis, 181
Broom, Percis Chapman, 72, 138, 155–56, 166
Broom, William, 72, 166
Brownson, Orestes, 154
Buchanan, James, 116
Bunyan, Paul, 2, 47, 179
Burnt House, 43–46, 47, 48
Butz, Earl, 186

Calhoun, Creighton Lee, 193
camp meetings, 106, 107, 108, 109, 123, 156
Canada, 94

canal lands, Indiana, 153, 155, 166
canals, 141, 150, 151, 153, 160, 163; Erie, 141; Miami and Erie, 151, 153; Ohio and Erie, 150; Wabash and Erie, 153
Captain Pipe, 94–95
Carter, Robert, 114
celibacy, 104–5, 109, 116–17, 124, 155
Chapman, Abner (brother), 70, 72
Chapman, Davis (brother), 72, 163
Chapman, Dorothy Swain, 13, 16
Chapman, Edward, 7, 9–18, 34
Chapman, Elizabeth (sister), 18, 21, 23, 27, 28, 34
Chapman, Elizabeth Simonds (mother), 18, 21–23
Chapman, John: and alcohol, 62, 147, 189; ancestry of, 7, 9–18; and apple seeds, 36, 40–41, 82–83; and bare feet, 36–37, 96, 98, 102, 120–21, 126, 127, 156; benevolence of, 57, 60, 119, 162, 168–69, 171, 181; celibacy of, 116–17, 124; as colporteur, 102–3; and crossing of Alleghenies, 36–38; feminine traits of, 2, 117–18, 189; frugality of, 25, 119–21, 164, 170–72; and grafting, 58–59, 149, 157; and land speculation, 130–35; and Leominster, 20–23, 34, 192; and Longmeadow, 27–31, 32, 34–35; and markets, 65, 123–24, 128, 161–62, 153, 181–82; meekness of, 2, 63, 64, 95, 99, 117–18, 168, 190; and Native Americans, 46–48, 91, 93, 153, 171, 173, 177–78; and New Church, 102–3, 106–7, 109–12; night run of, to Mt. Vernon, 96–97; and northwestern Pennsylvania, 36–40, 50; pacifism of, 96, 99, 180; and primitive Christianity, 103, 115, 119, 120, 124–27, 133, 147; and seedling nurseries, 42, 50, 59, 82–83, 85, 87, 139, 142–43; and Swedenborg, 103–5, 117, 119–120; as thriftless, 57–58, 60, 64; as vegetarian, 1, 49, 62, 96, 112, 114, 122–23; and War of 1812, 95–99; as Yankee, 1, 28, 36, 53, 137–38, 168;
Chapman, Jonathan Cooley (brother), 72, 163
Chapman, Lucy Cooley (stepmother), 26–27, 31, 32, 67, 68, 69, 72
Chapman, Nathaniel (Edward's son), 15–16
Chapman, Nathaniel, Jr. (brother), 27, 59, 70–72, 163, 166
Chapman, Nathaniel, Sr. (father): departure of, from army, 25–26, 32, 33–34; in Leominster, 18, 20; in Ohio, 68, 69–70, 71, 72; in the Revolution, 20–27, 54
Chapman, Parley (brother), 70, 72, 163

charity, 119, 128, 172, 181; and the Grange, 170; and Puritans, 11, 13, 15; and Swedenborg, 104, 119
Chesebrough, Silas, 133–34, 135
Child, Lydia Maria, 122, 171–72, 174
Chillicothe, Ohio, 69, 76–77
Church of the New Jerusalem: and John Chapman, 102–3, 106–7, 109–12, 117, 119–20; and Cincinnati, 112–14; in England, 102, 105, 112; and Philadelphia, 105–7, 110–12; and socialism, 113–14; and the West, 107–9, 105–15, 117–21, 123–24, 143, 168, 188; and Yellow Springs colony, 113–14
cider, 10, 19, 144; and brandy, 74, 143, 145, 147, 158, 191; and Harrison campaign, 158–59, 161; production of, 19–20, 32, 143; and temperance movement, 144–48, 158–59, 191
cider mills, 19, 40–41, 58, 64, 73, 75, 146–47, 191
Cincinnati, 69, 78, 111–14, 141, 142, 176
circuit riders, 108, 110, 116, 144
Civilian Conservation Corps, 178
Clear Fork, Ohio, 84, 90
Clowes, John, 102, 105, 106
Cobbett, William, 159
Coffinbury, Salathiel, 121–22
Cold War, 2–3, 178–82
colporteurs, 102–3
competency, 12–14, 17–18, 34, 54, 62, 85, 190
"conjugial" love, Doctrine of, 104–5, 117
Connecticut Valley, 31–32, 35, 67, 68, 81, 135, 149
Cooley family, 26–28, 30, 68–72, 78, 153
Copus, James, 95, 97–99
"Copus massacre," 97–99
Cornplanter, 24, 43–46, 47, 48, 60
Correspondences, Doctrine of, 105
Coshocton, Ohio, 84, 149–50, 163
Cowherd, William, 112
Craig, Andy, 85
Crèvecoeur, J. Hector St. John de, 49
Crockett, Davy, 2, 3, 49, 63, 173, 179, 190
Cuming, Fortescue, 93, 101–2
Curtis, R. I., 63–64, 121, 122, 173

Daniels, John, store ledger of, 48, 50, 58, 107, 129
Davenport, Stewart, 128
Dawes, Carlos Burr, 181, 188
Dawson, John, 119, 147, 156
deforestation, 47, 48, 64

Delaware (Lenni Lenape): and Greentown, 68, 90–93, 95–97, 99; and Sandusky, 152–53, 173
Detroit, 94, 193
Diamond, David, 189, 190
Donation Tract (Ohio Company Lands), 68–70, 71
Dow, Lorenzo, 156
Dowdee, Billy, 153
Downing, Andrew Jackson, 167, 175
Duck Creek Valley, Ohio, 71–72, 163
dwarf-stock apple trees, 182, 184, 188
Dwight, Timothy, 79–80

Edwards, Jonathan, 30
Ely, Nathaniel Jr., 32–33
Ensign, Silas, 110
environment, impact of white settlement on, 10, 47, 48, 64, 92–93, 101
environmentalists, 184–85, 193
Environmental Working Group, 193
evangelism: through printed word, 102–3, 106–7; through revivals, 106–9, 123, 156

Fairfield County, Ohio, 78, 81–82
Fallen Fruit, 193
Fallen Timbers, Battle of, 43, 69, 93
farmers, improving, 81–82, 128, 140–41, 146, 148
farmers, self-provisioning, 80–81, 134–35, 143, 150, 158–59, 191; and apples, 6, 11, 88–89, 148–49
Fast, Howard, 178–79
Fink, Mike, 2, 63
Finley, Alexander, 138–39
First Bank of the United States, 129
Fort Industry, Treaty of, 90
Fort Michilimackinac, 94
Fort Wayne, Indiana, 152, 155–56, 162, 164–65
Franklin, Benjamin, 106
Franklin, Pennsylvania (fort and village), 41–42, 45, 52, 61, 65
freemen, 12–13
French Creek, Pennsylvania, 38, 39, 41, 42, 57–58, 60–61, 63
Friends of Sholan Farms, 192
frontier: clothing on, 89–90; as decivilizing force, 48–49, 57–58; and market economy, 79–83; settlement process for, 79–80; and stereotypes, 57–58

frontier legends: Boone, 2, 3, 49, 173; Bunyan, 2, 47, 179; Crockett, 2, 3, 49, 63, 173, 179, 190; Fink, 2, 63. *See also* Johnny Appleseed (myth)
frugality, 119–21, 122, 154, 162, 164, 169, 170–72
Fruitlands, 124, 155. *See also* utopian communities
Fruits and Fruit Trees of America (Downing), 167, 175
Funk Bottoms Wildlife Area, Ohio, 138–39

Germans, 77, 97
Gibbs, Grant, 192
Gillette Family, 82, 175
Glen, James, 105–6, 107
Glines, W. M., 163–64, 173
Golden Delicious apple, 184–85
Gold's Gym, 187
Goldthwait, Johnny, 79, 81, 83
government certificates, 25, 31–32
Graham, Sylvester W., 112, 123
Grange, 169–70
Great Awakening (First), 18–19, 30
Great Black Swamp, 150–52, 153, 162
Greentown (Ohio Indian Village), 90–93, 95–97, 99, 153
Greenville Treaty Line, 43, 69, 83, 84
Gross, Lisa, 193–94
Grouseland (estate), 77, 158

Haley, W. D., 142–43, 169–71, 173, 174
Handsome Lake, 44, 92, 104
Harper's Monthly Magazine, 87, 98, 125, 142–43, 169–70
Harrison, William Henry, 77, 83, 91, 93, 158–59, 161
Hatch, Isaac and Minerva, 138
Hatch Act, 176
Heaven and Hell (Swedenborg), 106
Henry, John, 179
hogs, 6, 9, 75, 80, 101, 122, 133; and apples, 40, 146, 191
Holland Land Company, 45, 51–52, 55, 56, 58–62
honeybees, 92, 101
House Un-American Activities Committee, 178–79
Howe, Henry, 116, 118, 165, 168
Howells, William Cooper, 159–60
Hull, William, 94–95, 96
Humrickhouse, T. S., 149–50, 165, 168

hunting, 7, 9, 44, 47–49
Hurdus, Adam, 112–13
Huron River, Ohio, 91, 95–96
husbandry, mixed, 9–11, 43, 44, 47–49

Ikerd, John, 192
Indians. *See* Native Americans
Indian trails, 38–40, 43, 83
inheritance, partible, 14
IOUs, 31, 103, 129
Ipswich, Massachusetts, 7–15, 17, 18
Iroquois, 5, 17, 24–25, 29, 39, 41, 44–46
Irvine, Callender, 53–57
Irvine, William, 51, 52, 53–55, 61

Jackson, Andrew, 153, 154, 158, 177
James, John, 139–40, 143
Jefferson, Thomas, 25, 48, 76–77, 82, 91, 186
Jeffersonian Democrats, 81–82
Jerome's Fork, Ohio, 84, 90
Jerometown (Ohio Indian village), 90, 91, 92, 95, 99
Johnacake, Solomon, 153
Johnny Appleseed (myth): and animals, 1, 96, 122, 171, 180; bare feet of, 1, 36–37, 96, 98, 102, 120–21, 126, 127, 156; and Cold War, 2–3, 178–82; extent of apple tree planting by, 2, 3, 79, 86, 89, 189; frugality of, 28, 154, 169–72, 192; and Native Americans, 1, 46, 91, 153, 171, 173, 177–78; nonviolence of, 1–3, 178; religious faith of, 1, 179–81; as vegetarian, 1, 49, 62, 96, 112, 114, 122–23; as Victorian era creation, 2; and Walt Disney Company, 2, 179–81
"Johnny Appleseed Grace," 179
Johnny Appleseed Heritage Center and Outdoor Drama, 188
Johnny Appleseed: Man and Myth (Price), 26, 181, 188
Jones, Levi, 96–97
Jones, Robert, 132
Jones, William Ellery, 188–89

Kingsbury, Noel, 89
Kinmont, Alexander, 113, 139
Kinzua Reservoir, Pennsylvania, 44
Knight, Thomas Andrew, 140
Knox County, Ohio, 85, 115, 150
Kokosing Valley. *See* Owl Creek Valley, Ohio

land companies, 51–52; Holland Land Company, 45, 51–52, 55, 56, 58–62; Ohio Company and Associates, 68–69, 71, 73, 78, 81; Population Land Company, 51
land leases, 23, 13–32, 138, 140, 151, 166
land ownership: and competency, 12–14, 17–18, 34, 54, 62, 85, 190; improvement requirement for, 51, 53, 68, 131–33; and independence, 12, 14–17, 18, 34–35, 42, 49–50, 80, 190; and Ohio, 85; and partible inheritance, 14, 17; and private property, 49; and social standing, 11, 35
Leatherwood God, 104
ledger books, 31, 47, 59, 88–89, 129
Leominster, Massachusetts, 17, 18–23, 34, 191, 193
Licking Creek, Ohio. *See* Licking Valley, Ohio
Licking Valley, Ohio, 78–79, 81, 83, 171
Lincoln, Abraham, 108, 177
Lindsay, Vachel, 177
Longmeadow, Massachusetts, 26–35, 38, 68, 70–71, 81
low-oxygen storage facilities, 182, 184, 185, 186

Malus coronia. *See* apple trees: North American crab
Malus domestica. *See* apple trees: Old World
Malus pumila. *See* apple trees: Old World
mammon, 128, 136, 145, 147, 161
Manchester, England, 101–2, 105, 112, 126
Manchester Society for the Printing, Publishing, and Circulating of the Writings of Emanuel Swedenborg, 102–3, 105
manhood, 2, 18, 48–49, 65
Mansfield, Ohio, 84, 90, 98, 115, 120, 124, 142, 152, 162, 170
markets: and agricultural improvement, 81–82, 128, 140–41, 146, 148; and apples, 81–82, 158, 159–61, 176, 185–87, 192; and John Chapman, 65, 123–24, 128, 153, 161–62, 181–82; and frontier, 79–83; and materialism, 3, 123–25, 135–36, 155; and morality, 123–25, 145–46
marriage: and the frontier, 60, 62–63; and heaven, 23, 104–5, 117–18; and independence, 7, 12, 18, 27, 60, 62; as obstacle to perfection, 155; as unnatural state, 104–5, 116
Massie, Nathaniel, 76, 77
Maumee River, 92, 151, 153
Means, Howard, 189, 190

mediums of exchange: barter, 31, 40, 59, 88–89, 110, 129; bills of exchange, 70–71, 166; government certificates, 25, 31–32; IOUs, 31, 103, 129; ledger books, 31, 47, 59, 88–89, 129; paper money, 88, 100, 110, 129–33; promissory notes, 70–71, 88, 129; specie, 88, 110, 129–30, 153–54
Metcalfe, William, 112, 114, 123
Methodism, 81, 106–10, 113, 116, 124, 156
M'Gaw, James, 169
microhistory, 3–4
migration: to Ohio, 73–75; Pennsylvania Dutch, 76; Scots-Irish, 76; upland Southerner, 67, 76; to the West, 66–67; Yankee, 68
missionaries, 107
Mohekan John's Creek, Ohio, 84, 86, 88–89, 101, 102, 104
Mohican River, Ohio. *See* Mohekan John's Creek, Ohio
money. *See* mediums of exchange
Morgan, Robert, 189–90
Mumford, Lewis, 178
Muskingum River, Ohio, 68, 72, 75, 76, 79, 81, 83, 163
Muskingum River Valley, Ohio, 65, 83, 85, 86, 130

National Resources Defense Council (NRDC), 184–85
National Road, 162–63
Native Americans: adoption of orchards by, 24–25; and John Chapman, 1, 46–48, 91, 93, 153, 171, 173, 177–78; Greentown, 46–48, 90, 95–97; mixed subsistence strategies of, 7–8, 44–47, 49, 83; in New England, 7–9, 12; in Owl Creek region, 83, 90, 91; Sullivan's raid on, 24–25; and War of 1812, 94, 95, 97–99; western confederacy of, in 1790s, 41–44, 69; and whites, 45–46, 91; white stereotypes of, 25, 173–74. *See also* Delaware; Iroquois; Sagamore Indians; Seneca Indians; Shawnee; Wendat
navel orange, 176
Nessly, Jacob, 74, 143
New Church. *See* Church of the New Jerusalem
New Harmony, Indiana, 113–14
Newport, Thomas, 107
Nicholson, John, 51
Northwest Ordinance, 76, 131
nurseries: John Chapman's described, 79, 86, 87, 103, 171; Goldthwait's, 81–82; Putnam's, 81, 82, 89; squatter, 59, 75, 79, 85–86, 88, 138–40; Van Horn's, 41–42, 64

Ohio Company and Associates, 68–69, 71, 73, 78, 81
Ohio Cultivator, 167–68
Old Southern Apples (Calhoun), 193
olive trees, 10
On Atonement (Ballou), 115
orange trees, 10, 176
orchards: decentralized, 19, 32, 193–94; grafted, 5, 69, 73, 82, 143, 148, 160–61, 175; Native American, 24–25, 41; seedling, 5, 20, 80, 82, 85, 143; urban, 192–94
Owen, Robert, 113
Owl Creek Bank, 130–34
Owl Creek Valley, Ohio, 83–86, 100, 104

Palmer, Caleb, 95–96, 99, 128
Panic of 1819, 131, 134–35, 136, 138, 141
Panic of 1837, 154–55, 158, 167
paper money, 88, 100, 110, 129–33
Paris Green (ferrous sulfate), 175
patriarchy, 14–16, 17, 34–35
Patrons of Husbandry, 169–70
Payne, Adam, 156
Peace Corps, 3
peaches, 24, 25, 56, 68, 74–75, 77–78, 80, 160
peddlers, Yankee, 133, 135, 137–38, 190; John Chapman as, 137–38, 168
Pennsylvania Dutch, 76, 78–79, 168
Pennsylvania Land Act of 1792, 51–53, 57, 61–62, 64–65
perfectionism, 3, 123, 155, 192
Perry, Oliver Hazard, 99, 128
Philip Seymour (M'Gaw), 169
Pollan, Michael, 188–89, 190
pomace, 40–41, 147
Population Land Company, 51
Portland Fruit Tree Project, 193
poverty: and John Chapman, 88, 89, 120–21, 161–62; and Nathaniel Chapman family, 33–35, 67, 72, 190; and frontier, 51–52, 67, 73–74, 76–77, 80, 99; and Puritans, 8–9, 11; and seedling apple trees, 89, 137, 161, 191
Powell, David, 107
Price, Robert, 26, 181, 188, 190
private property, 49
promissory notes, 70–71, 88, 129

Puritans, 7–15; and Great Migration, 7; and land distribution, 11–13
Put-In Bay, Battle of, 99, 128
Putnam, Israel, 68, 81, 82, 89

Quaker missionaries, 48, 60

railroads, 175–76
Red Delicious apples, 4, 182–84, 189, 191
refrigeration, 175–76
religion: in England, 108–9; in Ipswich, 17th-century, 7–15; in Leominster, 18th-century, 18–19, 30–31; in Longmeadow, 18th-century, 28–31, 35; Native American, 45, 48, 92; revivals of, 106, 107, 108, 109, 123, 156; and the West, 108, 109
revivals, 106, 107, 108, 109, 123, 156
Rhode Island Greening apples, 9–10, 68
Rice, Eben, 88, 102, 129
Rice, Rosella: as author of sentimental fiction, 169; descriptions of Broom family by, 138, 155–56; descriptions of John Chapman by, 116, 118, 121, 137–38, 157, 169–71; and Haley, 169–71, 174
Richland County, Ohio, 88, 140, 142, 150, 152, 162
Rocky Fork, Ohio, 84, 90, 96, 130
Roe, Daniel, 113
Rome Beauty apples, 82, 175
rootstock, 4–5, 66, 182, 184
Ruffner, Martin, 97, 99
Rush, Benjamin, 67, 79–80, 88, 144

Sagamore Indians, 7–8, 12, 17
"Sammy Appleseed," 168, 174
Sandy Creek, Pennsylvania, 39, 62
Schlatter, William, 106–7, 110–11, 112
Scots-Irish, 76, 168
Second Bank of the United States, 110, 154
Second Great Awakening, 3
self-provisioning farmers. *See* farmers, self-provisioning
Seneca Indians, 43, 50
Seven Ranges, 73–76, 77
Shakers, 109, 124
"sharp shins," 88
Shawnee, 90, 92–93, 95, 99
Shays's Rebellion, 31
"Sixtyfoot Smiths," 112
60 Minutes, 184–85

Slocum, Elias, 120–21
smallpox, 15, 17–18, 27
Smith, Adam, 128
southerners, 67, 76, 82, 86
specie, 88, 110, 129–30, 153–54
Specie Circular Act, 153–54
speculators, currency, 31–32
speculators, land, 51–53, 56, 61; and John Chapman, 130–35
Springfield, Massachusetts, 23–25, 33–35
Springfield arsenal, 23–26, 33
squatter(s), 51, 53, 54–55, 64–65, 76, 85; John Chapman as, 59, 86, 139, 140, 150, 190; Nathaniel Chapman as, 71–72
Stadden, Catherine, 78–79, 81, 83, 168, 171
Stadden, Isaac, 78–79, 81, 83
stealing, apple, 159–61, 167
Stemilt Growers, 184, 186
storage facilities, controlled atmosphere, 182, 184, 185, 186
Streep, Meryl, 184
Sturges, E. B., 152–53
subsistence lifestyle, mixed, 7–8, 43, 45–47, 48–49, 83
suckers (derogatory term), 67, 83, 88, 99. *See also* poverty
suckers (root sprouts), 66–67, 78, 175
Sullivan, John, 24–25, 38–39, 41, 44, 45
Swedenborg, Emanuel, 102–5, 117, 119, 142. *See also* Church of the New Jerusalem
Swedenborgians: Bailey, 106, 113, 139; Carter, 114; Clowes, 102, 105, 106; Cowherd, 112; description of, 114; Ensign, 110; Glen, 105–6, 107; Hurdus, 112–13; James, 139–40, 143; Kinmont, 113, 139; Metcalfe, 112, 114, 123; Newport, 107; Powell, 107; Roe, 113; Schlatter, 106–7, 110–11, 112; Sixtyfoot Smiths, 112; John Young, 106. *See also* Church of the New Jerusalem

Tall Hunter, The (Fast), 178–79
Tannehill, Melzar, 135, 137
Tannehill, Nancy, 116, 118, 135, 137
taxes, 31–33
Tecumseh, 93, 99
temperance, 20, 112, 143–48, 158–59, 191
Tenskwatawa, 92–93, 95, 101, 104
Thomas, Josiah, 121
Thoreau, Henry David, 5–6, 20, 124, 146–47, 177

thrift, 28, 119, 124, 168, 181
Tibbets, Eliza Lovell, 176
Tiffin, Edward, 76, 77, 80
Tippecanoe, Battle of, 93–94
Toby (Indian), 95–96, 99, 128
Tocqueville, Alexis de, 102, 123, 128
top-grafting, 74, 80, 82, 137, 148

ultraism, 55, 144, 145, 147
Uniroyal Chemical, 184–85
United States Department of Agriculture (USDA), 175–76, 186
universalism, 114–15
Urbana, Ohio, 139
Urbana University, 188
urban orchard movement: Boston Tree Party, 193–94; City Fruit, 193; Fallen Fruit, 193; Friends of Sholan Farms, 192; Portland Fruit Tree Project, 193
U.S. Apple, 186–87
utopian communities, 109, 113, 114, 124, 155

Van Buren, Martin, 158
Van Horn, Cornelius, 41, 42, 52, 60–64
Van Mons, Jean Baptiste, 149
vegetarianism, 112, 114, 122–24, 135, 154. *See also* Chapman, John: as vegetarian
Venango County, Pennsylvania, 61, 63
violence, in frontier myths, 2, 3, 173
Virginia colony, 10
Virginia elite, 76–77, 82, 131, 158
Virginia Military District, 76–78, 131; school lands, 131–32, 134, 138, 155, 166
Virginians, 67, 76, 77, 86

Walhonding River, Ohio. *See* White Woman's Creek
Walt Disney Company, 2–3, 179–80
War of 1812, 93–100
Warren, Pennsylvania, 36, 37, 43, 60
Washington Apple Growers, 187

Wayne, Anthony, 42
Wendat (Wyandot), 68, 90, 95
Wetmore, Lansing: 36–41, 43, 49–50, 120, 165; account of Chapman's Allegheny crossing by, 36–39
Whaley, Richard, 132
Whig Party, 81–82, 154, 158–59
White Woman's Creek, 84, 86, 100
Whitney, John, 163–64
Whitney, Sally Chapman, 72, 163–64
wild apples. *See* apple trees, seedling
wilderness: as decivilizing force; 48–49, 57–58; white attitudes toward, 47–49, 79–80, 93
Williams, Eunice, 29, 35
Williams, John, 73–75, 77, 78
Williams, Stephen, 28–30, 46
Winthrop, John, 73–75, 77, 78
Winthrop, John, Jr., 7, 8, 11
Worth, Richard, 166
Worth, William, 164
Worthington, Thomas, 76–77, 78, 80, 82, 131
Wyoming Valley, Pennsylvania, 24, 36, 38, 41, 61
Wyoming Valley Yankees, 53, 54–55

Yankee peddler(s), 133, 135, 137–38, 190; John Chapman as, 137–38, 168
Yankee-Pennamite War, 38
Yankees: John Chapman as, 1, 28, 36, 53, 138, 168; improvers, 142, 167, 194; migration to Ohio of, 68, 76, 86; squatters in northwest Pennsylvania, 53–55; stereotypical characteristics of, 1, 25, 28, 36, 135, 152, 168; and Wyoming Valley, 38, 53, 54–55
Yellow Springs, 113–14
Yellow Sweeting apples, 9–10
Young, John, 106
Young, Matt, 55–56

Zane's Trace, 75, 76, 78, 79, 81, 83
Zanesville, 69, 76, 79, 84, 142, 163
Zimmer, Philip, 97, 169